NATURAL RESOURCE MANAGEMENT

Volume 1

Genes from the Wild
Using Wild Genetic Resources for Food and Raw Materials

Full list of titles in the set
Natural Resource Management

Volume 1: Genes from the Wild
Volume 2: Waterlogged Wealth
Volume 3: The Threatening Desert
Volume 4: Our Common Seas
Volume 5: Asking the Earth
Volume 6: The Rural Economy and the British Countryside
Volume 7: The Water Crisis
Volume 8: After the Green Revolution
Volume 9: Saving the Seed
Volume 10: Unwelcome Harvest
Volume 11: The Politics of Industrial Agriculture
Volume 12: Forest Politics
Volume 13: No Timber Without Trees
Volume 14: Controlling Tropical Deforestation
Volume 15: Plantation Politics
Volume 16: Saving the Tropical Forests
Volume 17: Trees, People and Power
Volume 18: Tropical Deforestation

Genes from the Wild
Using Wild Genetic Resources for Food and
Raw Materials

Robert and Christine Prescott-Allen

First published in 1988 by
International Institute for Environment and Development

This edition first published in 2009 by Earthscan

Copyright © International Institute for Environment and Development 1983, 1988

For a full list of publications please contact:

Earthscan
2 Park Square, Milton Park, Abingdon, Oxon OX14 4RN
711 Third Avenue, New York, NY 10017, USA

First issued in paperback 2018

Earthscan is an imprint of the Taylor and Francis Group, an informa business

All rights reserved. No part of this book may be reprinted or reproduced or utilised in any form or by any electronic, mechanical, or other means, now known or hereafter invented, including photocopying and recording, or in any information storage or retrieval system, without permission in writing from the publishers.

Notice:
Product or corporate names may be trademarks or registered trademarks, and are used only for identification and explanation without intent to infringe.

A catalogue record for this book is available from the British Library

Library of Congress Cataloging-in-Publication Data has been applied for

Publisher's note
The publisher has made every effort to ensure the quality of this reprint, but points out that some imperfections in the original copies may be apparent.

At Earthscan we strive to minimize our environmental impacts and carbon footprint through reducing waste, recycling and offsetting our CO_2 emissions, including those created through publication of this book. For more details of our environmental policy, see www.earthscan.co.uk.

ISBN 13: 978-1-138-92880-0 (pbk)
ISBN 13: 978-1-84971-012-1 (hbk)

GENES FROM THE WILD

using
wild genetic resources for
food and raw materials

by Robert and Christine Prescott-Allen

Illustrations by Bryan Poole

EARTHSCAN PUBLICATIONS LTD
London

This edition published 1988 by
Earthscan Publications Limited
3 Endsleigh Street, London WC1H 0DD

Copyright © International Institute for
Environment and Development 1983, 1988

First published 1983 by the International Institute for
Environment and Development, London

Genes from the wild was produced with funding from
the Swedish International Development Authority (SIDA).
However, this paperback does not necessarily represent
the view of this or any other organisations.

Some of the data in this report were obtained in the
course of research funded by World Wildlife Fund–US
and Philip Morris Inc., whose support the authors
gratefully acknowledge.

Earthscan Publications Ltd is an editorially independent
and wholly owned subsidiary of the International Institute for
Environment and Development (IIED).

British Library Cataloguing in Publication Data:

Prescott-Allen, Robert
   Genes from the wild. —2nd ed.
   1. Crops. Genes. Variation
   I. Title II., Prescott-Allen, Christine, *1949–*
   631.5'3

ISBN 1-85383-026-7

Contents

1. **The oldest resource; the newest resource** **9**
   Some definitions 10

2. **What have wild genetic resources been used for?** **13**
   Cereals 14
   Root crops 20
   Oil crops 23
   Vegetables and pulses..27
   Fruits and nuts 29
   Sugar crops 31
   Commodity crops 34
   Fibre crops 36
   Timber 37
   Forage crops 39
   Livestock 43
   Aquaculture..47

3. **The nature of wild genetic resources** **49**
   Benefits of wild genetic resources 50
   What kinds of wild species are used?.............................54
   The future of wild genetic resources.............................54

4. **Where are wild genes found? And who uses them?** **60**
   Who has got them? And who benefits? 60

5. **Threats to wild genetic resources**......................... **71**
   Cereals 71
   Root crops 73
   Oil crops 74
   Vegetables and pulses..76
   Fruits and nuts 76
   Sugar crops 77
   Commodity crops 78
   Fibre crops 79

Timber 79
Forage crops 81
Livestock 83
Aquaculture 83

6. Conservation of wild genetic resources 85
In situ gene banks 89
The difficulties 92

References 95
Index 108

Executive summary

Utilisation of wildlife is as old as the human species, but the conscious use of wild genetic resources is a 20th century phenomenon. .... Page 9

Between 1930 and 1975, US crop yields rose dramatically due to use of genetic resources. .... Page 9

A crop's primary gene pool consists of those species that hybridise easily with it; its secondary and tertiary gene pools are species that can only be crossed with greater difficulty. .... Page 11

Crops have already used wild genetic resources extensively; livestock has not. Their use is described for each of the world's main crops: cereals, roots, sugars, trees etc. .... Page 13

Several wheat cultivars get resistance to fungal disease from wild wheats, but new disease strains usually evolve for most crops, to which new resistant genes must be found, especially from the wild. .... Page 14

Wild wheat also has genes for drought resistance, winter hardiness, heat tolerance. .... Page 16

Rice gets its resistance to two of Asia's four main rice diseases from a single sample of wild rice from central India; crops with its genes are now grown on 30 million hectares in Asia, and have doubled Indonesia's rice production. .... Page 16

In 1977 a new species of wild maize was found in Mexico, whose genes carry resistance to four serious diseases. .... Page 18

Until 1851 Europe's potatoes were all descended from two 16th century samples: a genetic bottleneck. After the 1845/6 Irish potato famine, new wild genes came from South America. .... Page 20

Genes from wild cassava have increased yields in Africa and India by up to 18 times. .... Page 22

Southeast Asia's oil palm crop was descended from only four palms taken to Java in 1848, and yields were gradually raised over threefold by breeding from this material. Since 1957 new plantings have been based on new wild genetic material from Africa. .... Page 24

Most vegetables do not rely on wild genetic resources, but the tomato could not be grown as a commercial crop without them. .... Page 27

Wild peas and beans contain many valuable characteristics, but in spite of breeders' efforts these have not yet been transferred to the crop. ... Page 28

Without wild sugarcane genes, there would probably not be a viable sugarcane industry anywhere in the world. .... Page 32

Wild cotton has provided a strain which does not produce the leaf nectar that attracts insect pests, thus reducing the need for pesticides. .... Page 36

Tree crops take from 10 to 80 years to mature, so breeding is a longer process. But an FAO programme has completed collection of material from 20 priority wild species. .... Page 38

Australian scientists recently found genes in Colombia and Mexico which could extend a legume pasture crop to drier, colder districts. ...... Page 40

In Israel, a new hybrid goat has been bred from wild ibex, and in the US the "beefalo" is a cross between cattle and bison. .... Page 43

Wild bee genes have been little used, but some silkworms in India are being improved with wild germplasm. .... Page 46

In the USSR, a cross with a wild fish has enabled carp production to be extended to colder regions. Hybrid *Tilapia* in Africa give vigorous all-male hybrids that yield well, but only if the parents come from Zanzibar and Lake Albert respectively. .... Page 47

Few major crops have not been improved with wild genes, and the frequency of their use is increasing, especially for disease resistance. Breeders go to wild relatives of crops only as a last resort, because unwanted genes often come with desirable ones. .... Page 49

Genetic engineering and prebreeding is being increasingly used. ... Page 56

There is a wide belief that the South conserves genetic resources for the North to exploit. This is not wholly true: not all the useful genes are found in the South. .... Page 60

Wild genes move mainly North-North and South-South, but in the South-North direction the North is at present a clear gainer. . . . . Page 62

Gene exchange has been based on free movement and leaving duplicates in the country of origin. But these principles are breaking down, and FAO is preparing an international convention. Opinions differ on its likely value. . . . . Page 67

Although the South now has a growing gene bank and plant breeding capacity, long term storage facilities are mainly in the North. But by 1985 this will have changed. . . . . Page 68

Patenting of genes poses new problems, as does the ability to transfer desirable genes to bacteria and "grow" the crop in a laboratory. . . . . Page 69

Wild genes, like wild species, are threatened by habitat loss, over-exploitation and introduced species. But risks are harder to detect. . . . . Page 71

The main crops are reviewed to see how much their wild genes are threatened. In cereals, some wild grains are benefiting from new habitats created by urban expansion, but most are declining. . . . . Page 71

Several wild sunflowers are threatened in California, one by the growth of Los Angeles. Wild oil palms may be threatened by deforestation in Africa, and are definitely endangered in Latin America. . . . . Page 74

Oil exploration in Ecuador threatens the centre of diversity of wild cacao trees. . . . . Page 79

Wild timber genetic resources are threatened. A wild cypress species in Algeria is down to 153 trees. . . . . Page 79

Many wild cattle, horses, goats, sheep and pigs are vulnerable or endangered, but most wild relatives of chickens, turkeys and ducks are not. . . . . Page 83

Wild genetic resources can be conserved *ex situ* (in zoos and gene banks) and *in situ* (in nature reserves). . . . . Page 85

Conservation (*in situ* or *ex situ*) may involve base collections, active collections, or working collections. Base collections are for long term security of genes; working collections are for breeders to use. . . . . Page 85

The IBPGR has promoted *ex situ* base collections for over 20 key crops. Thirty-three long term seed banks now exist compared to only seven in 1975. However, the bulk of the material is cultivated germplasm . . . . Page 85

Of 38 rice gene banks, only five provide long term storage for wild rice; of 53 maize gene banks, only one stores wild maize genes long term. .... Page 87

Legume collections contain more wild material, but there is almost no storage of wild coffee from Madagascar, of potential since its beans are very low in caffeine. .... Page 88

In situ gene banks are more hope than reality. Four genuine ones are known to exist: in the USSR (2), India and Sri Lanka. Nature reserves rarely even document their wild genetic resources. .... Page 89

National parks and (even more so) biosphere reserves are two philosophies that could be extended to include wild germplasm. .... Page 92

Both *in situ* and *ex situ* genetic resources conservation are necessary; *in situ* conservation is now the priority. .... Page 94

1. The oldest resource; the newest resource

Wild genetic resources combine the oldest resource, wildlife, and the newest resource, genes. The use of wild plants and animals is as old as the human species. The use of genes is a phenomenon of the 20th century.

People have long taken advantage of genetic variation, when they have selected plants or animals with characteristics they prefer: plants with bigger seedheads; hens that lay more eggs; cows that give more milk. But genes were not consciously used as a resource until we knew they existed.

Austrian monk Gregor Mendel laid the cornerstone of the science of genetics in 1866, when he published the results of seven years work with more than 30,000 pea plants; but his findings had no impact until they were rediscovered in 1900. The science that then flowered so rapidly was first called "genetics" in 1906; the term "gene" was coined in 1909 (120).

The use of genetic resources in agriculture has been the most far-reaching application of the new science. In the 45 years between 1930 and 1975, US yields per hectare of wheat rose by 115%. Over the same period, US rice yields rose by 117%, of maize by 320%, of sugarcane by 141%, of peanuts by 295%, of soybeans by 112%, of cotton by 188% and of potatoes by 311% (128). About 50% of each of these increases can be attributed to the contribution of genetic resources (131).

Similarly, the average milk yield of cows in the United States has more than doubled over the past 30 years. Genetic improvement accounts for more than 25% of this increase, at least with respect to Holsteins, one of the breeds participating in the US Dairy Herd Improvement Program (131).

These phenomenal improvements, where they have been due to the application of genetics, have largely been obtained using the genetic resources of the *domesticated* species concerned. As yet, the contribution of *wild* genetic resources has been comparatively small.

Domesticated genetic resources, which were as much of an untapped resource prior to the turn of the century as were wild genetic resources, are much easier to work with. Nonetheless, as this report shows, some crops (such as sugarcane) have been transformed by genes from their wild relatives. Other crops (such as tomatoes) probably could not be grown on the current scale without them, and the domestication of still others (such as timber and forage crops) has been revolutionised through the use of wild genetic resources.

These achievements are substantial enough to demonstrate the enormous potential of the genetic diversity of wild plants and animals, to improve the yields and quality of domesticated crops and livestock, and to provide for the more rapid domestication of new crops and livestock.

It is an attractive prospect. The use of wild genetic resources is a means of benefiting materially from wildlife without harming the donor species. Obtaining these benefits, however, depends partly on a more efficient use of wild genetic resources. But it depends much more on maintaining as great a range as possible of the potentially useful genetic variation within wild species. This in turn requires a clearer understanding of the special nature of wild genetic resources: how they differ from wildlife and other wild resources, and from domesticated genetic resources.

Some definitions

Genetic resources are actually or potentially useful characteristics of plants, animals and other organisms, that are transmitted genetically. So *wild genetic resources* are any heritable characteristics of a wild plant or animal that are of actual or potential use to people. The characteristic may be rapid growth, disease resistance, a chemical property, an environmental adaptation, or the capacity of a tree to grow tall and straight. As long as it is or is likely to be of economic or social value, is found in a wild species, and is transmitted genetically, it qualifies as a wild genetic resource.

The term wild genetic resource is sometimes used interchangeably with species diversity and genetic diversity, as if they all have identical meanings. *Species diversity* means the variety of different species, while *genetic diversity* means the variety of genes. Genetic diversity is normally used to cover diversity *within* species, while species diversity is the term for diversity *among* species.

Genetic resources is a category of genetic diversity, referring to the within-species variation that has been, is being or is likely to be used in the selection or improvement of domesticates, or the manipulation and enhancement of wild stocks.

Genetic resources are often qualified in various ways. *Crop genetic resources* are genetic resources used in agriculture; *forest genetic resources* are genetic resources used in silviculture and *aquatic genetic resources* are genetic resources used in aquaculture. And, of course, *domesticated genetic resources* are the genetic resources of domesticated plants and animals, and *wild genetic resources* are the genetic resources of wild plants and animals.

Wild genetic resources differ from their domesticated counterparts chiefly in terms of conservation, which we discuss in Chapters 5 and 6. The crucial distinction is the wildness of the plants and animals concerned. Just what do we mean by wild? A *wild plant or animal* is one that reproduces independently of human control. It is not sown or planted, put to stud, or given the assistance of a vet at delivery; but it propagates itself. Its critical habitats, those required for reproduction and nutrition, can regenerate without human intervention. Wild species are adapted to closed, primary habitats; to open, naturally disturbed habitats; or to range or forest disturbed by human activity. They are not wild if they survive only or largely in cultivated fields and gardens or urban areas.

The likelihood that a wild genetic resource will be used depends on the importance of the plant or animal to be improved, the rarity of the genetic characteristic (is it the only source of resistance to a disease, or one of many?), and the ease with which the characteristic can be transferred to the domesticate.

The term *gene pool* means the total number of genes within a group of inter-breeding plants or animals; that is, the pool of genes within a population. Two University of Illinois professors, Dr Jack Harlan and Dr Jan de Wet, have given the term a second meaning (71): the total number of genes within a domesticate and its wild relatives; that is, the pool of genes that is potentially available for the improvement of the domesticated plant or animal. They divide this gene pool into three categories: primary, secondary, and tertiary, depending on the ease of gene exchange.

The *primary gene pool* (GP1) consists of the domesticated species plus those wild forms that are inter-fertile and hybridise readily with it. The primary gene pool corresponds to the traditional concept of the biological species. Within it hybrids are usually fertile, and gene transfer is generally easy.

The *secondary gene pool* (GP2) consists of those biological species that can be crossed with the domesticated species using conventional breeding methods to produce at least some fertile progeny. Gene transfer is possible but less easy, as many hybrids are sterile or difficult to bring to maturity.

The *tertiary gene pool* (GP3) consists of those species that can be crossed with the domesticate, but from which gene transfer is possible only through the use of special techniques, such as embryo culture, grafting or tissue culture. Hybrids are usually completely sterile or non-viable.

Harlan and de Wet stress that their system is informal. It should not be applied rigorously; GP3 in particular is a highly flexible concept, defining "the extreme outer limit of potential genetic reach" (68), which is almost certain to expand with advances in genetic engineering.

The tomato provides an example of how the Harlan-de Wet system can be applied. GP1 consists of *Lycopersicon esculentum*, which comprises both the domesticated tomato and the wild and weedy variety *cerasiforme*, plus *L. pimpinellifolium* and *L. cheesmanii*, two wild tomato species that are completely inter-fertile with *L. esculentum*.

GP2 consists of three more tomato species: *L. chmielewskii*, *L. hirsutum* and *L. parviflorum*, plus a species in the potato genus *Solanum pennellii*, all of which can be crossed with the domesticated tomato with varying degrees of difficulty.

GP3 at present consists of the remaining two species in the *Lycopersicon* genus (*L. chilense* and *L. peruvianum*), plus two species of *Solanum*, *S. lycopersicoides* and the potato itself *S. tuberosum*. *L. peruvianum*, for example, can be crossed with the tomato only with the help of embryo culture (145; 146a). This example shows clearly that the taxonomic classification of species is not a sure guide to their genetic relationships.

Inside the nucleus of every animal or plant cell are *chromosomes*, thread-shaped bodies consisting largely of DNA (deoxyribonucleic acid). Short lengths of each chromosome are called *genes*, and genes confer particular characteristics on the organisms that inherit them.

Each gene is in effect a chemical instruction controlling a particular characteristic. There is a gene for eye colour in humans, for example. One variant of that gene gives blue eyes, another gives brown, another green. Such variants of the same gene are called *alleles*.

Differences in the genetic make-up of a species are caused by different alleles of each gene. In the oil palm, for example, the shell of the kernel of the fruit is controlled by a gene *sh*, one allele of which (*sh+*) gives a thick shell while the other allele (*sh-*) gives no shell at all.

Like many plants and almost all animals, oil palms are diploids, which means that their chromosomes and their genes occur in pairs. If the alleles of the shell gene are both *sh+*, the palm has fruit with thick shells; if they are both *sh-*, it has fruit with no shells; if one allele is *sh+* and the other *sh-*, the palm has fruit with thin shells. An oil palm with either thick shells or no shells is said to be *homozygous*, because both alleles are the same. If the palm has thin shells, it is *heterozygous*, since the alleles of the same gene are different.

A particular combination of genes is known as a *genotype*. A given set of chromosomes is called a *genome*.

2. What have wild genetic resources been used for?

This chapter discusses the contribution of wild genetic resources to the production of food and raw materials. We have restricted our examples, as far as possible, to crop and livestock resources of world importance.

Crop and livestock groups vary greatly in the use they have made of wild genetic resources. In general, crops have already been considerably improved using wild genetic resources, while livestock have not. Most food crops have been domesticated for some time, and they need wild genetic resources for particular kinds of improvement. Most timber trees, forage crops (food for animals) and fish, on the other hand, are today only just on the threshold of domestication. If scientific, this process involves the selection of the best genotypes from the best gene pools of the species being domesticated, but, if traditional, it does not. The domestication of forage crops has been largely scientific, making increasing use of genetic variation within species; while that of aquacultural livestock (fish, oysters etc) has been largely traditional, using whatever material is most readily to hand.

There are also great differences within crop and livestock groups. Of the commodity crops, for example, cacao has benefited most from its wild relatives, in terms both of the number of improvements given and the proportion of the crop improved. Tobacco is the next biggest beneficiary, a large proportion of the crop getting resistance to several major diseases from wild species.

But in coffee and rubber, genes from wild species have not yet had a commercial impact, although there is a well-recognised need for them, and it is only a matter of time before they do make a contribution. Finally, tea has considerable unexploited variation within the crop itself, and so appears to have little need of wild genetic resources at least for the foreseeable future.

In reviewing each crop or livestock group we have tried to make a clear distinction between those plants and animals that have already been improved using wild genetic resources, and those for which use of wild germplasm is still a twinkle in a breeder's eye. We also describe some of the difficulties breeders encounter when working with wild species. The contributions of wild genetic resources are remarkable given some of the problems of using them, and the time it can take to develop a new, commercially acceptable breed or cultivar (cultivated variety).

Our sample of cereals, root crops, vegetables and pulses, fruits and nuts, sugar crops, commodity crops, fibre crops and livestock resources, is taken from the *FAO Production Yearbook* (46); and of oil crops from *Agricultural Statistics* (174) of the United States Department of Agriculture. For each of these groups we give the world average annual production for the three-year period 1978-80; and for each crop we give the percentage of group production that it accounts for. For example, world production of commodity crops is 18 million tonnes, of which cacao provides 9%. We concentrate on the crops that account for 10% or more of world production in each group and have excluded crops that account for less than 1%.

Production figures provide only a very rough indication of the relative importance of these resources at the world level, and no indication at all of nutritional or economic importance to particular countries. Equivalent figures are not available for forage crops, timber crops or aquaculture resources.

Where known and relevant we indicate the crop gene pool to which each wild species belongs: primary gene pool (GP1), secondary gene pool (GP2) or tertiary gene pool (GP3). For definitions of these terms, see Chapter 1.

Cereals (1,575 million tonnes per year)
Four crops – wheat, rice, maize and barley – together make up almost 90% of the world's annual production of grain, and all have been improved with wild genetic resources.

Wheat (28%)

The two most widely grown forms of wheat are the common bread wheat (*Triticum aestivum vulgare*) and the macaroni wheat (the *durum* form of *T. turgidum turgidum*). Wild species have not been used successfully to improve macaroni wheat, although Dr Lydia Avivi of the Weizmann Institute of Science in Israel has a programme to raise its protein content by transferring genes for high grain protein from the wild form *T. turgidum dicoccoides* (GP1) (52a).

Several commercial cultivars (varieties) of bread wheat, however, get valuable characteristics including their resistance to strains of fungal diseases from distantly related wild species.

* In the USSR the high-yielding cultivar Dneprovskaya 521 was developed from a cross with a wild species of rye (*Secale*) (GP3); and the frost resistance of the cultivar Orenburgskyi 40 comes from a wild wheatgrass (*Agropyron*) (GP3) (21).
* The French wheat cultivar Roazon has a gene (from *Aegilops ventricosa:* GP3) which gives it resistance to *Cercosporella herpotrichoides*, cause of eye spot, a disease of the leaves just above the soil line (122a).
* A number of US and Australian cultivars (for example, Agent and its derivatives in the US and Eagle, Kite, Avocet and Jabiru in Australia)

WHEAT. Genes from the wild goatgrass Aegilops ventricosa *(left) provide wheat with resistance to the leaf disease eye spot, and the crop-plant einkorn* Triticum monococcum *(centre) is a source of rust resistance. The most widely grown form of bread wheat,* T. aestivum vulgare, *is on the right.*

have at least partial resistance to leaf rust and stem rust from *Agropyron elongatum* (GP3); *Aegilops umbellulata* (GP3) is another source of leaf rust resistance in the US (137; 122a).

Stem, leaf and stripe rusts, caused by *Puccinia* fungi, are the most serious diseases of wheat. In epidemics they have destroyed up to 75% of the crop, and even in non-epidemic years annual losses to stem rust alone can exceed 4% or 2.3 million tonnes in the US (111). The most effective means of controlling rusts is through the breeding of resistant varieties, so the genetic contributions of the wild species have been important.

Disease resistance that is controlled by a single dominant gene, as opposed to several minor genes, seldom lasts. Sooner rather than later new races of the pathogen (the disease organism) evolve to which the "resistant" cultivars prove susceptible. The new races of disease do not overcome the resistance of the crop; the crop genes, being resistant only to particular races of disease, are simply ineffective against new ones. Consequently, the breeding of disease resistant cultivars is a continuous process.

If resistance is available in the primary gene pool, so much the better, since obtaining it from the secondary — and especially the tertiary — gene pools can be extremely laborious. The transfer of leaf rust resistance from *Aegilops umbellulata* to wheat, for example, required first the use of the wild *Triticum turgidum dicoccoides* (GP2) as a bridging species, and then irradiation of the progeny to translocate the resistant gene from the *umbellulata* chromosome to a wheat chromosome (103). Irradiation was also necessary in the case of *Agropyron elongatum* (103).

Wheat breeders are fortunate that they have available to them a variety of sources of rust resistance — in the crop species itself, in domesticated species in the secondary gene pool (einkorn *T. monococcum*, emmer and macaroni wheat *T. turgidum*) and the tertiary gene pool (rye *Secale cereale*), as well as in other wild species besides those already used. The domesticated relatives have in fact been the major supply of resistance genes so far.

Despite the difficulties of working with the more distantly related wild species, wheat breeders continue to do so in the belief that wild genes are the resource of the future. One reason is that many plant diseases, particularly the stem rust and leaf rust fungi, have a remarkable capacity to evolve new races in response to new conditions. So breeders will need every source of resistance at their disposal.

Disease resistance is not the only potential contribution of wheat's wild relatives. They are repositories of drought resistance, winter hardiness, heat tolerance, salt tolerance, earlier ripening, higher productivity and increased protein content. The wheat breeders Moshe Feldman and Ernest R Sears (the latter was responsible for the first commercial use of a wild wheat species) contend that the more distantly related wild species are more likely to be useful than the close relatives. "In spite of the difficulties of transferring genes to wheat from its more distant relatives", they write, "attempts to make such transfers are decidedly worthwhile. The greater the distance over which transfers can be made, the greater the possibility of introducing useful characteristics not present in the cultivated wheats" (52). The theory is that closely related species, which might occasionally hybridise naturally, already share a pool of common characteristics. Brand new characteristics must come from outside that pool.

Rice (25%)

Rice is a simpler story than wheat. There has been only one successful use of a wild species, but it has had a bigger impact. The crop gets its resistance to blast and grassy stunt virus — two of the four major diseases of rice in Asia (the other two are bacterial blight and tungro virus) — from the wild rice species *Oryza nivara* (GP1).

During the early 1970s, before the release of resistant cultivars in 1974, grassy stunt epidemics destroyed more than 116,000 hectares (287,000 acres) of rice in Indonesia, India, Sri Lanka, Vietnam and the Philippines. Today, with the widespread use of resistant varieties, the disease has ceased to exist in farmers' fields.

Cultivars with *O. nivara* in their pedigree are now grown on 30 million hectares (74 million acres) in India, Nepal, Bangladesh, China and Southeast Asia. Their economic and social impacts have been enormous. In Indonesia, average production per hectare has risen to six tonnes and in some cases to 10 tonnes. Two to three crops a year have become common, and rice production overall has doubled. Rice availability per person in Indonesia has increased from 91 kg (200 lbs) in the 1960s to 136 kg (300 lbs) today (16).

A curious aspect of the GSV (grassy stunt virus) resistance of *O. nivara* is its extreme rarity. Professor Gurdev Khush, who bred the resistant varieties at the International Rice Research Institute (IRRI) in the Philippines, and the late Dr KC Ling, the pathologist who discovered the resistance, screened 6,723 accessions (samples) of cultivated and wild rice species. Only one accession of *O. nivara* was found to be highly resistant to GSV (100). Subsequently Dr Ling did discover a second potential source of GSV resistance, in a cross between rice and another wild species in the crop's primary gene pool, *O. rufipogon*. However, when we visited the IRRI, shortly before Dr Ling died in 1982, he had not yet succeeded in multiplying enough seed to test the resistance (110a).

Why is GSV resistance so rare? One reason is that the disease was a minor one until the development of high-yield cultivars that could be grown year round. This provided both a permanent reservoir for the disease, and a regular source of abundant food for the brown planthopper (*Nilaparvata lugens*), the insect which spreads GSV. In the absence of these conditions the disease was so rare that few farmers knew it existed.

To many people, including prominent scientists in the plant breeding community, it was an act of faith if not of weakmindedness to conserve a wild rice like *O. nivara* in the IRRI gene banks. The plant has no obvious useful characteristics, and many that are highly undesirable. It is squat and floppy with weak stems; it has a very low yield; and like most wild cereals its seedhead shatters at maturity, scattering the seeds on the ground. When it was collected in 1966, GSV was an unimportant disease, so no samples of rice were being checked for resistance. If botanist Dr SD Sharma had not collected this single resistant sample of wild rice from central India, and sent it to the IRRI in the Philippines, this vital characteristic might never have been discovered. Such is the lottery of agricultural improvement.

Maize (25%)

All of the improvement in maize (*Zea mays*) has been achieved using the genetic resources of the crop itself, with the exception of two cultivars, Texas 42SX and Texas 30A, which contain some wild genetic material. These two were reported to be among the most popular hybrids in the southwest US and northern Mexico during the late 1960s and early 1970s, growing on an estimated 100,000-200,000 hectares (247,000-494,000 acres) a year. They had higher-than-average yields and an absence of top firing (killing of upper leaves during hot weather) and chlorophyll breakdown. According to their breeder Dr AJ Bockholt, these characteristics were due to genes from the wild gama grass *Tripsacum dactyloides* (GP2) (137). Texas 42SX and Texas 30A have since been replaced by hybrids that mature earlier.

The *Tripascum* gama grasses have been dismissed by one botanist as "essentially worthless", but another wild species offers great promise for the future. In 1977 the Mexican botanist Rafael Guzman made two exciting discoveries. He refound *Zea perennis*, a wild relative of maize previously thought to be extinct. And he made what has been hailed as the "botanical find of the century": the discovery of *Zea diploperennis*. *Z. perennis* is a tetraploid (has four sets of chromosomes), however, and can be crossed with maize only with great difficulty (GP2). But *Z. diploperennis*, like maize itself, is a diploid (two sets of chromosomes) and hybridises readily (GP1). So it is a potential genetic resource of great value.

As their scientific names suggest, both these wild corns are perennial. This means that one plant grows for a number of years; cultivated maize is annual, and so has to be re-sown each year. Since the discovery of diploperennial corn, there has been considerable ballyhoo about the prospect of transferring the perennial character of the wild species to the crop. Visions have been painted of permanent stands of maize, like orchards of fruit or plantations of trees. It would be extremely rash to declare that this will never happen, but it does seem unlikely that a perennial maize will be developed, except possibly as a forage crop. Perennial grasses have major disadvantages. They are poor and erratic setters of seed, because they do not depend on seed production for survival (68). And perennials tend to be less productive than annuals because they must share their nutrient supplies between the seeds and the vegetative structures that enable them to overwinter (129).

A much more interesting characteristic of the wild maize *Z. diploperennis* is its unusual disease resistance. It is immune to four serious diseases of maize:
* Maize chlorotic dwarf virus: *One of the two most serious viral diseases in the US. No other source of immunity is known.*
* Maize chlorotic mottle virus: *This disease is serious in South America as well as in parts of the US. No other source of immunity is known.*
* Maize streak virus: *This virus causes the most serious viral disease of maize in Africa. No other source of immunity is known.*
* Maize bushy stunt mycoplasma: *A serious disease at high elevations in tropical Central and South America.*

Wild maize *Z. diploperennis* is also tolerant of two other viral diseases (maize stripe and maize rayado fino), and is tolerant or immune to Strain B of maize dwarf mosaic virus (129). Even if only 0.25%-1% of the US maize crop were to be saved from these diseases, the saving would be worth an estimated $50-250 million a year (93a).

Barley (11%)

The wild barley *Hordeum spontaneum* (GP1) has been used in Europe as a source of resistance to *Erysiphe graminis*, the cause of powdery mildew, probably the most important fungal disease of cultivated barley (*H. vulgare*). Some extremely effective systemic fungicides are now available to control powdery mildew, but the use of resistant cultivars is "probably the main control method and . . . likely to remain so" (154). This wild barley has been widely used in European spring barley breeding programmes, notably in West Germany (Akme was the first commercial cultivar with this material) and in the United Kingdom (for example, the cultivars Maris Badger and Maris Concord released in the early 1960s) (186).

Powdery mildew, like wheat rusts, evolves very rapidly, and the resistance is ineffective against new races of the fungus that have appeared. So the popularity of varieties with the resistance gene from *H. spontaneum* has declined (154). However, the species shows promise in other directions, such as improving the productivity of the crop by increasing grain yield (176).

BARLEY. Wild barley Hordeum spontaneum *(left) has been used in Europe as a source of resistance to powdery mildew, probably the most important fungal disease of cultivated barley (right).*

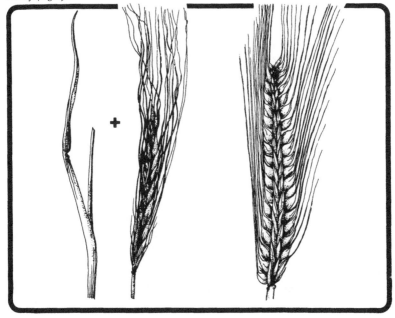

Sorghum (4%), Oats (3%), Millets (2%), Rye (2%)

As far as we know, oats (*Avena sativa*) is the only one of these crops that has been improved with wild germplasm. Genes from wild oats (*Avena sterilis:* GP1) have been used to provide resistance to rust diseases (137).

Root crops (522 million tonnes per year)
Potato, cassava and sweet potato together make up 94% of the world root crop. All three root species gain disease resistance from wild genetic material.

Potatoes (50%)

The potato is among the first crops to have benefited from genes from a wild relative. When cultivated potatoes (*Solanum tuberosum*) were taken to Europe from South America they went through a genetic bottleneck. Until 1851 the European crop seems to have been based entirely on only two introductions, one to Spain in about 1570 and the other to England in about 1590 (76). This meant that only a tiny segment of the genetic diversity of the South American cultivated species was represented in Europe, and this genetic uniformity made the European crop extremely vulnerable to epidemics.

The genetic vulnerability of the potato was demonstrated on a tragic scale in 1845 and 1846, when the crop was severely attacked by late blight, caused by the fungus *Phytophthora infestans*. Epidemics occurred throughout northern Europe, but Belgium and Ireland were hardest hit. In Ireland, especially, the impact of the disease was most terrible. The potato was the staple food of the mainly peasant population, and loss of the crop led to widespread famine. Death from starvation, combined with emigration to Britain or North America, reduced the population of Ireland from 8.2 million in 1841 to 6.2 million in 1851 (154).

With such a limited array of genetic material, the early potato breeders found themselves like card players with only a tiny part of the deck. From 1851 new introductions were made from South America, and although none was resistant to late blight they did provide some valuable genetic variation. The first breakthrough came in 1908, when the British plant breeder RN Salaman showed that the Mexican species *Solanum demissum* and its natural hybrid with the potato *S. edinese* were resistant to late blight. The second came in 1925, when the Russian plant breeder NI Vavilov, who conceived the idea of making detailed collections throughout the range of a cultivated species, sent his colleague SM Bukasov on the first scientific expedition to collect wild and domesticated potatoes in Middle and South America (75; 76).

Today, out of 586 potato cultivars grown in Europe (West and East, excluding the USSR), 320 have genes from wild species (calculated from Reference 166). Out of 71 potato cultivars grown in the Soviet Union, 30

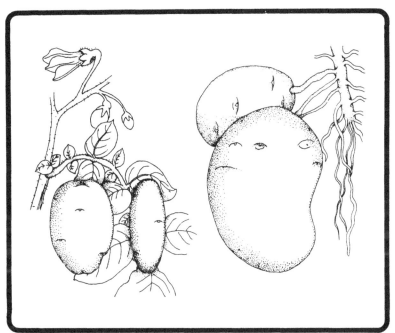

*POTATO. Cultivated potatoes in Europe (*Solanum tuberosum: *right) were for nearly 300 years descended from only two introductions from South America. After the Irish potato famine of 1845/6, caused by the fungus disease late blight, genes from the wild Mexican potato* S. demissum *(left) provided resistance.*

contain genes from the wild (152). *Solanum acaule* (GP2) provides resistance to two virus diseases, potato virus X and potato leaf roll virus. *S. spegazzinii* (GP2) confers resistance to five races of the cyst nematode *Globodera rostochiensis* as well as to the fungus *Fusarium coeruleum*. *S. stoloniferum* (GP2) has provided several cultivars, particularly in the Netherlands and Federal Germany, with extreme resistance to potato virus Y. These two countries have also made extensive use of *Solanum vernei* (GP2), for its resistance to some races of the cyst nematode *Globodera pallida*. *S. vernei* has also conferred the unexpected benefit of high starch content (152).

The most widely used wild potato species, however, is *S. demissum* (GP2). Originally used for its major gene resistance to late blight, breeding with this aim in mind was largely abandoned following an outbreak in 1936 of races of late blight against which this type of resistance was ineffective. But cultivars with *S. demissum* in their background were later found often to be more resistant to late blight than cultivars without, due to the actions of minor genes. These "polygenes" confer a measure of resistance to all races of late blight which, while by no means complete, is a good deal better than nothing. In addition, cultivars with *S. demissum* are resistant to potato leaf roll virus, this resistance also being polygenically controlled (151; 152; 61a). Field resistance to late blight can be increased by crossing potato cultivars containing *S. demissum* germplasm with each other, so accumulating the polygenes (151). Hybrid vigour and increased potential yield are apparently other benefits of incorporating *S. demissum* material (152).

Cassava (23%)

In recent years great progress has been made, at the International Institute of Tropical Agriculture (IITA) in Nigeria, in developing improved cassava cultivars with resistance to cassava mosaic disease and cassava bacterial blight. The two diseases are the most serious diseases of cassava in Africa. No

CASSAVA. The wild cassava Manihot glaziovii *(left) is the only known source of resistance to the two most serious cassava diseases in Africa. Transfer of its genes to cultivated cassava (right) has increased yields of new cultivars by up to 18 times.*

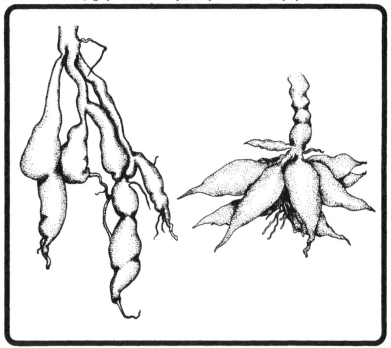

adequate source of resistance has yet been found in the cultivated crop (*Manihot esculenta*) (64; 65), but it has been discovered in the wild *M. glaziovii* (GP2), whose resistance to both diseases is apparently controlled by a complex of minor genes (63).

This gene complex has been successfully transferred to the new cultivars, enabling them to yield from two to 18 times more than traditional varieties in Nigeria, Sierra Leone, Liberia, Gabon, Zaire, Tanzania, the Seychelles and India (65). Breeders at IITA have also used *M. glaziovii* to lower the cyanide content of cassava, and are turning to other wild species, such as *M. tristis saxicola* (GP1) to raise the protein content (65a). Cassava is also known as manioc or tapioca.

Sweet potato (21%)

Sweet potatoes (*Ipomoea batatas*) are not related to true potatoes; originating in South America, they are now grown mostly in south and east Asia.

Minamiyutaka, a Japanese cultivar, gets its resistance to root-lesion nematodes (*Pratylenchus* spp) and root-knot nematodes (*Meliodogyne* spp) from the wild *Ipomoea trifida* (GP1) (157). Since its release in 1975, the area grown to Minamiyutaka has risen rapidly, from 0.5% of Japan's sweet potato area in 1976 to 5.7% in 1980 (157a; 157b).

Oil crops (39 million tonnes oil equivalent per year)
Soybean, sunflower and oil palm together provide over half the world's plant oil. Although soybean has used little wild genetic material, sunflower and oil palm provide graphic examples of the value of wild genes.

Soybean (31%)

In the USSR, the wild *Glycine soja* (GP1) has been used in the soybean (*G. max*) breeding programme of the All-Union Soya Research Institute, with the aim of obtaining cold tolerance in the germination and seedling stages and early ripening (21). We do not know if this has led to successful cultivars. Otherwise no germplasm from wild species has been used in the improvement of this crop (137).

Sunflower (13%)

The sunflower (*Helianthus annuus*) owes its position as the world's no. 2 oil crop to two major achievements in plant breeding: the development of high oil yielding cultivars in the USSR, and of hybrid lines in the US using a source of cytoplasmic male sterility discovered in France.

Soviet scientists raised the oil content of sunflower seeds from 20-30% in

1915, to 55% in 1965, by breeding within the crop (140). They also turned to the cultivated Jerusalem artichoke (*Helianthus tuberosus*) (GP2) for resistance to the parasitic plant broomrape (*Orobanche cumana*), and to some of the wild species, for resistance to several fungal diseases.

But a far bigger contribution of wild species has been their role in the development of hybrid cultivars. The advantage of hybrids is their much increased productivity as a result of the vigour conferred by combining two different homozygous lines; in the case of sunflower the potential yield increase is 25% greater than that of cultivars pollinated in an uncontrolled way. To obtain hybrid seed, the breeder must ensure that line A is pollinated only by line B, preventing individuals in line A from either pollinating themselves or being pollinated by other line A individuals. This can be done by emasculating (cutting out the pollen-bearing organ) the members of line A, but this is an extremely expensive, labour-intensive operation. A simpler and far cheaper method is to ensure that the pollen of line A is sterile.

In cytoplasmic male sterility (CMS), the gene for male (pollen) sterility occurs not on a chromosome but in the cytoplasm, the main contents of a cell. The phenomenon has been found in 80 species, 25 genera and six families of plants, so it is quite widespread; but stable CMS that is 100% effective under a wide range of conditions is rare (74). Sources of genetic fertility restoration (GFR) — which restores fertility to the hybrid seed that the farmer actually grows as the crop — are more common. A combination of CMS and GFR prevents self-pollination but ensures that seed given to farmers will grow flowers and seed.

A stable source of CMS was discovered by Patrice Leclercq in France from a cross between the cultivated sunflower *H. annuus* and the wild sunflower *H. petiolaris* (GP1). The next year, in 1970, the American breeder Murray Kinman reported a source of fertility restoration (GFR) in a wild form of *H. annuus*. Hybrids using the *H. petiolaris* source of CMS are now grown on 90% of the North American sunflower area and are widely grown elsewhere in the world. Although genes for fertility restoration have been found in domesticated sunflowers, they are more common in the wild species which provide most of the sources of GFR in current breeding programmes. The hybrids are generally more disease resistant (with resistance to some diseases also obtained from the wild) than previous cultivars; they mature uniformly (open-pollinated crops ripen at different times); and they yield from 18% to 25% more than their open-pollinated predecessors (137).

Oil palm (12%)

When oil palm (*Elaeis guineensis*) was taken from its native West Africa to Southeast Asia, it went through an even more severe genetic bottleneck than did the potato when it went from South America to Europe. The crop was established in Malaysia and Indonesia from the progeny of only four

SUNFLOWER. *Hybrids using genes from wild sunflowers (*Helianthus petiolaris: *left) are now grown on 90% of the North American sunflower area, and have helped the cultivated sunflower (*H. annuus: *right) become the world's No 2 oil crop.*

individual palms introduced into the botanical garden at Bogor, Java, in 1848. The first breeding programme from this material, made around 1920, raised yields from 1.2 tonnes of oil per hectare to 2 tonnes per hectare (0.5-0.9 tons per acre), or just over 65%. Subsequent generations of selection increased yields by a further 35% and improvement in plantation practices added another increase of 50%, raising average oil yield per hectare to almost four tonnes (two tons per acre).

It was clear to oil palm breeders in the two countries that no further improvement could reasonably be expected from such a narrow genetic base. So they returned to Africa for new germplasm from the wild. Since 1957, new plantings in Malaysia and Indonesia have consisted largely of crosses between the original thick-shelled fruit type (called *dura*) and a type (called *pisifera*) with no shell at all. Among the hybrids of the two, a third form appears (called *tenera*) whose shell is intermediate in thickness. Since 1967 in Malaysia and 1971 in Indonesia, all new plantings have been of *tenera* material obtained by crossing the original *duras* with wild *pisiferas*, mostly from Zaire but also from the Ivory Coast and Nigeria. The result has been a further increase in oil yield of more than 25%, so that the average yield is now five tonnes per hectare (2.6 tons per acre) (137).

Other oil crops

Of the other major oil crops, maize (1% of world production) has been discussed under cereals, and cottonseed (8%) will be covered in the section on fibre crops. Among the remainder — peanut (8%), rapeseed (8%), coconut (8%), olive (4%), sesame (2%), linseed (2%), safflower (1%) and castor (1%) — only sesame has been successfully improved with germplasm from the wild. A high yield cultivar has been developed in India from a cross between a southern Indian traditional variety and *Sesamum orientale* var. *malabricum* (GP1), the wild gingelly of Malabar (11a).

---

*TOMATOES. The tomato (*Lycopersicon esculentum: *right) probably could not be grown as a commercial crop without wild genes. Today, almost all varieties get their resistance to* Fusarium *wilt from the wild Peruvean tomato* L. pimpinellifolium *(left).*

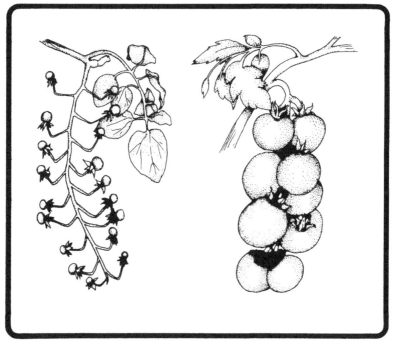

Vegetables and pulses (394 million tonnes per year)
Although wild tomatoes have been crucial to the commercial crop, and carrots and bell peppers have benefited from wild genes, other vegetables have so far been improved without the use of wild genetic resources.

Tomatoes (13%)

Few crops have benefited more from their wild relatives than has the tomato (*Lycopersicon esculentum*). According to Dr Jack Harlan (69), the tomato simply could not be grown as a commercial crop without the genetic support of its wild relatives. It gets almost all of its disease resistance from wild species, as well as several improvements in quality: better nutritional content, more intense colour, higher content of soluble solids and adaptation to mechanical harvesting. Those same wild relatives have also provided characteristics that enable tomatoes to be grown and eaten year round, even if culinary romantics dismiss them as superior only in "gassability and crashworthiness" (Thomas Whiteside in the *New Yorker*) and as an "almost total gastronomic loss" (James Beard in *Beard on Food*).

The unusually comprehensive contribution of wild species to the tomato can perhaps be attributed to its atypical origins. The three wild species in its primary gene pool (the original wild tomato, *Lycopersicon esculentum* var. *cerasiforme*, plus the two related species *L. pimpinellifolium* and *L. cheesmanii*) are restricted to Ecuador and Peru. The four species in the secondary gene pool (*L. chmieleswkii*, *L. hirsutum*, *L. parviflorum* and *S. pennellii*) have, if anything, a still more restricted distribution since *L. hirsutum* is found in Ecuador and Peru and the other three species occur only in Peru (37; 126; 147).

The wild form (*cerasiforme*) of the cultivated tomato is believed to have spread as a camp follower from Ecuador north to Central America and eventually to Mexico, where it was domesticated. This was the first genetic bottleneck. It went through other genetic bottlenecks when subsequently the crop was transported to other parts of the world. The wild tomato *L. esculentum cerasiforme* spread with the crop, and is now widespread as a weed, but outside its home in Ecuador and Peru it is also rather uniform genetically (145).

The crop that has grown to be the world's most important vegetable, and a major item in the food and commerce of such countries as Egypt, Mexico, Brazil, the US, China, Turkey, Greece, Italy, Romania and Spain, has done so from an extremely narrow genetic base. Given the relative lack of variation in the domesticated species, it is not surprising that breeders have resorted so extensively to the various wild related species.

Until 1941, tomato cultivars had little resistance to disease. That year, the first cultivar with resistance to the wilt-causing fungus *Fusarium oxysporum* was released in the US. Its resistance was obtained from *L. pimpinellifolium* (135), and today almost all cultivars have resistance to *Fusarium* wilt from this source (99a). Unlike the rusts of wheat or the mildews of barley, which are airborne diseases, *Fusarium* is soil-borne and new races spread more slowly. So the resistance genes have proved effective for quite long periods. A second race of *Fusarium* did not become a problem until the early 1960s, and has been successfully controlled with a second gene from *L. pimpinellifolium* (154). Altogether, genes from wild species provide the tomato with resistance to a dozen diseases and one pest, the root-knot nematode (*Meloidogyne incognita*) (146).

Many improvements in quality have been contributed from wild species to the tomato. They include:
* more intense internal and external colour from four wild species, including, rather remarkably, the greenfruited species *L. hirsutum* and *L. chmielewskii* (134; 146);
* increased soluble solids from *L. chmielewskii* (134; 146);
* higher vitamin C from *L. peruvianum* (173);
* higher beta-carotene (provitamin A) from *L. hirsutum* (146);
* a change to the pedicel (stalk of the individual fruit) that adapts it to mechanical harvesting, from *L. cheesmanii* (146);
* and, also from *L. cheesmanii*, an unusually thick skin (146): the characteristic that adapts the modern tomato so well to the rigours of transportation.

Pulses

The pulses include beans (chiefly *Phaseolus* and *Vigna* species, collectively making up about 4% of world vegetable and pulse production), peas (4%), broad beans (2%), and chickpeas (2%), as well as a number of minor crops such as lentils. None, as far as we know, has yet been improved with genes from wild species, except for peas in the USSR using *Pisum fulvum* (GP2).

This is not for want of trying. Wild beans and peas contain many valuable characteristics, but breeders have had great difficulty in making crosses, or incorporating the wanted characteristics, or in recovering the desirable features of the domesticate in the progeny of the crosses.

For example, the most economically important disease of chickpeas in North Africa and West Asia is blight caused by the fungus *Ascochyta rabiei*. It seriously reduces yields of the spring-sown crop, and completely prevents the sowing of a winter crop (78). Resistance has been found in cultivars of the *desi* type (the small, dark-coloured form preferred in India) but not in those of the *kabuli* (large and cream-coloured, and the only form grown in North Africa and West Asia). Even better resistance has been found in the wild *Cicer reticulatum* (GP1), but it has been difficult to transfer this resistance to the domesticate (78; 117). Resistance to another major disease of chickpea wilt caused by the fungus *Fusarium oxysporum* has been found in *C. judaicum* (GP3), but this has also not been successfully crossed (117a).

Similar problems are faced by breeders working with the *Phaseolus* beans. Most efforts to transfer valuable characteristics (such as disease resistance, cold resistance, resistance to lodging (falling over), and vigorous root systems) from wild forms of the runner bean (*P. coccineus*) to its GP2 relative *P. vulgaris* (the species that gives us baked beans, black beans, and snap, green or French beans) have led to frustration caused by sterility and incompatibility (59a).

Breeders face an even more extreme situation with the broad bean (*Vicia faba*). Broad beans are susceptible to a large number of diseases to which few good sources of resistance have been discovered. The wild progenitor of the crop species has not been found (it may be extinct), and the domesticate alone forms the primary gene pool. There is no secondary gene pool. One of the closest wild relatives and one of the few possible candidates for the tertiary gene pool is *V. narbonensis*, resistant to some of the crop's main diseases and pests, but all attempts to cross it with the broad bean have been unsuccessful (78).

Other vegetables

Other major vegetables include cabbages (10%) watermelons (6%), onions (5%), cucumbers and gherkins (3%), carrots (3%), chilis and bell peppers (2%), melons (2%), squashes (1%), eggplants (1%) and garlic (1%). Wild forms have been used to improve only carrots and bell peppers as far as we know.

Hybrid carrots are produced in the US using a source of male sterility from Queen Anne's lace, the wild form of carrot *Daucus carota* (GP1) (120a), a common weed in North America and Europe. Disease resistance from wild *Capsicum annuum* (GP1) has been transferred to bell peppers in the US, France and Argentina (103a; 134a; 137).

Reports that wild species have been used in the US to provide disease resistance in watermelons and melons in fact refer to escaped domesticated species (137). It is possible that wild germplasm has been used in watermelon breeding in Taiwan but we have been unable to verify this. Considerable work is being done with the wild relatives of other vegetable crops — such as onions, cucumbers, and squashes — but commercial cultivars have not yet been released.

Fruits and nuts (282 million tonnes per year)
Grapes rely heavily on rootstocks developed from wild American vines. Strawberries and several minor fruits have drawn extensively on wild germplasm, and banana breeders have turned to wild stocks for disease resistance.

Grapes (23%)

French grape breeders in the 19th century imported American vines in an attempt to obtain resistance to the powdery mildew fungus *Uncinula necator*. But with the vines came a far more serious problem: the plant-sucking insect pest *Phylloxera vitifoliae*. Its impact was even more disastrous than that of late blight on the potato. Between 1870 and 1900, virtually all the French vineyards were destroyed, and by 1910 so were most of the vineyards in Spain and many in Germany, Italy, Austria, Romania, USSR and elsewhere (3).

The answer to the problem: bring in more vines from North America, with resistance to *Phylloxera*. Grapes, like most garden roses and many fruit trees, consist of a fruiting plant (the *scion*) grafted onto a root (the *rootstock*) of another variety or species. Today, all grapes in *Phylloxera*-contaminated areas are grown on rootstocks that come from combinations of three North American wild vines — *Vitis berlandieri, V. riparia* and *V. rupestris* (all GP1).

These and other North American species, such as *V. labrusca, V. lincecumii* and *V. aestivalis* (all GP1), have also been crossed directly with the grape (*V. vinifera*) to form new scions. The hybrid cultivars are acceptable sources of grapes for grape juice, brandy and plonk but are not highly regarded as a source of superior wine grapes. The proportion of France's vineyard area grown to them has declined from 31% in 1958 to less than 20% in 1978 (60). The North American species remain the sole source of rootstocks. Disease resistance is the most valuable characteristic of wild *Vitis* but by no means the only one. In the USSR breeders have developed cold-resistant, winter hardy grape cultivars using *Vitis amurensis* (GP1), a wild grape native to East Asia (62).

Bananas (13%) and plantains (7%)

Cultivars incorporating wild germplasm have not yet been successfully developed for bananas and plantains (which are essentially the same, although dessert fruits are usually called bananas, and those used for cooking are usually called plantains). But they are urgently needed. The Cavendish clones, which are the basis of the banana export industry, are extremely susceptible to disease. Disease resistance has been found in wild bananas (*Musa acuminata*) (GP1), and has been transferred to breeding lines. A 15-year programme of crossing and selection has already yielded hybrids but it is estimated that another 10 years work is needed to combine this disease resistance with enough agronomic excellence for a commercial cultivar (153: 153a).

Citrus fruits

Oranges (13%), the tangerine group (includes mandarins, clementines and satsumas: 3%), lemons and limes (2%), grapefruit and pomelo (2%) and other citrus fruits are highly variable and cross freely. There is disagreement among *Citrus* specialists as to which species are truly wild. As far as we can tell, there is little if any deliberate use of wild germplasm in *Citrus* improvement.

Apples (12%)

Several *Malus* species have been used to provide disease and pest resistance and to widen the adaptations of the cultivated apple *M. pumila*. In the US, Canada and France new fruiting cultivars with genes from *M. floribunda* (GP1) (105) have been released with resistance to apple scab, which is caused by the fungus *Venturia inaequalis* and is probably the most serious disease of the apple wherever it is grown (20). In the United Kingdom, rootstock cultivars are being developed, with resistance to the mildew-causing fungus *Podosphaera leucotricha*, from *M. baccata* and *M. zumi* (both GP1) (2; 2a) In the USSR and China, breeders have used the cold-tolerance of *M. baccata* to extend the northward range of apple growing (39).

We are not entirely convinced of the wildness of some of these species. Apple breeders, like other tree-fruit breeders, often use the terms "*Malus* species", "small-fruited species" and "wild" interchangeably to refer to species other than that of the crop they are working with. Several *Malus* species are grown as ornamentals, however, just as are several species of *Prunus* (cherry, peach, plum etc) and *Pyrus* (pear). Although such species remain small-fruited, a number have undergone selection for their flowers and are true domesticates. *Malus floribunda*, the Japanese crab apple, is one such widely planted species.

Other fruits and nuts

Among the other major fruits, breeding of mango (5%) and dates (3%) does not appear to be sufficiently advanced to require the use of wild relatives: there is still considerable scope for selection within the crop. A variety of valuable characteristics (high acid and sugar content, good flavour, hardiness, drought tolerance, disease and pest resistance) have been transferred from wild species to the pineapple (3%), but no commercial cultivars have resulted.

Wild species have had a major impact on strawberries (1%) and on some of the minor fruits such as raspberries and currants. Immunity to scab in pears (3%) has been obtained from a wild species in the Soviet Union (21). Wild species have had a negligible impact so far on peaches and nectarines (2%), plums (2%) and apricots (1%), although many non-commercial cultivars of plums have been developed in North America using wild species. Deliberate use of wild material has had no discernible impact on papayas (1%) or nuts, of which the main ones (almonds, walnuts, chestnuts, filberts, cashews and pistachios) collectively make up about 1% of total world fruit and nut production.

Sugar crops (1,023 million tonnes per year)
Both cane and beet sugar breeders have used wild germplasm, extensively in the case of sugarcane.

Sugarcane (73%)

Sugarcane (*Saccharum officinarum*) has been transformed by its wild relatives. The vigour and disease resistance obtained from wild *S. spontaneum* (GP2) and to a much lesser extent from *S. robustum* (GP1) have had a major impact on world sugarcane production, helping to almost double the cane yield and to more than double the yield of sugar (161). The wild species have physically transformed the crop, so that commercial cultivars no longer have the expected appearance of sugarcane. The thick, rich-looking canes of the *officinarum* type have given way to thin, scrawny types, the products of half a century of breeding to combine the sugar quality of *officinarum* and the vigour, high potential yield and disease and pest resistance both of the wild relatives and of the less distinguished but hardier domesticated *sinense* types from India.

Although they may not look impressive, the new sugarcanes have performed superbly. They have rescued sugarcane in areas such as Louisiana in the US from almost certain extinction. They have enabled sugarcane industries to be established in countries — India, for example — which could not otherwise have grown the crop commercially.

The limiting factor to sugarcane production in India used to be the disease red rot. Then resistance was acquired from wild Indonesian *S. spontaneum*, which also provides resistance to two other devastating diseases of sugarcane: smut (*Ustilago scitaminea*) and sugarcane mosaic virus. Indonesian *spontaneum* genes are present in many of the most important commercial cultivars in the top sugarcane producing countries: South Africa, Cuba, Mexico, the US, Argentina, Brazil, Colombia, China, India, Indonesia, Pakistan, Philippines, Thailand and Australia. Probably equally widespread are *S. spontaneum* clones from India, which are sources of resistance to gummosis caused by *Xanthomonas vasculorum* bacteria and to root rot caused by the fungus *Pythium arrhenomanes*.

The wild sugarcane *S. spontaneum* has a wide range, from Africa to the western Pacific, but sources other than Indonesia and India have not been used until recently because of difficulties in synchronising the flowering of the wild and the cultivated clones. This problem has now been solved, and the latest cultivars have additional characteristics from new sources of *S. spontaneum*, such as adaptation to high elevations obtained from wild cane populations from Thailand.

Another wild species, *S. robustum* (GP1), has so far proved to be of more limited value. It has been used only in Hawaii, for vigour and stalk thickness, where it is also being tried out as a source of resistance to rats: the surface of its cane is hard and very difficult for a rat to gnaw (137).

The overall significance of the wild species for world sugarcane production has been summed up for us by the US sugarcane scientist, Dr JD Miller (121a): "If no germplasm from wild relatives had been used, there would most probably not be a viable sugarcane industry any place in the world".

Sugar beet (27%)

Sugar beet production has benefited from wild germplasm to a very much lesser extent. Wild sea beet *Beta maritima* (GP1) has contributed durable resistance to leaf spot caused by the fungus *Cercospora beticola*, a disease which is particularly severe in southern Europe (especially Italy and Spain), eastern Europe, Turkey and parts of North America.

For some time, breeders have been attempting to transfer resistance to the sugar beet nematode worm (*Heterodera schachtii*), perhaps the major pest of the crop, from the wild *B. procumbens* (GP2). This and two other species in the crop's secondary gene pool, *B. patellaris* and *B. webbiana*, are the only known sources of high resistance to the pest. In 25 years of work, breeders have had the greatest difficulty in ridding the few successful crosses of undesirable characteristics, without also getting rid of the gene for nematode resistance. Recently a satisfactory breeding line has been developed and commercial cultivars with nematode resistance should be released in the not too distant future (137).

SUGARCANE. Vigour and disease resistance obtained from wild sugarcane Saccharum spontaneum *(left) have helped more than double the sugar yield of cultivated sugarcane (*S. officinarum: *right), allowing it to be grown commercially in India and elsewhere.*

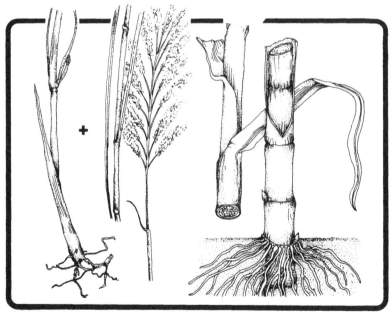

Commodity crops (18 million tonnes per year)
Wild germplasm has been widely and successfully used in both tobacco and cacao, but not at all in tea. Coffee and rubber are two crops with a narrow genetic base, and breeders are now using wild material, but not so far with commercial success.

Tobacco (31%)

Wild species from South America and Australia have made substantial contributions to the disease resistance of tobacco. They are the only sources of resistance to angular leaf spot (caused by *Pseudomonas angulata* bacteria), and to "wildfire", which is caused by *Pseudomonas tabaci* bacteria and so called because it spreads like wildfire (43). Both diseases occur in all the major tobacco-growing areas of the world. Seed dressings, sprays and cultural practices provide only partial control, and growers rely on resistant cultivars. The wild tobacco *Nicotiana longiflora* (GP2) has probably been used most widely, providing resistance to cultivars in the US, Zimbabwe, Japan and elsewhere. Several other wild species show promise. For example, tobacco breeders in Armenia, USSR, where wildfire is particularly serious, are using *N. plumbaginifolia* (GP2) (154). Other major diseases currently controlled by genes from wild species are tobacco mosaic virus (by *N. glutinosa:* GP2), black shank (by *N. longiflora* and *N. plumbaginifolia*), and black root rot (by *N. debneyi:* GP2) (137).

Coffee (27%)

The two main domesticated species of coffee are *Coffea arabica*, which produces "arabica" coffee, and *C. canephora*, which produces the inferior "robusta" coffee. Wild *C. arabica* occurs only in Ethiopia and southeastern Sudan, and was taken by Arabs to the Yemen, where it was domesticated. Eventually, a limited amount of this germplasm was dispersed to the current coffee-growing countries of the world.

Outside Ethiopia, the Sudan and the Yemen, cultivated coffee therefore has a very narrow genetic base. Efforts to widen this base are being made in a number of countries. The most urgent need is to obtain better resistance to the fungus *Hemileia vastatrix*, which causes coffee rust, the most serious disease of coffee. The disease was first identified in Sri Lanka in 1869, and in the following two decades it caused such widespread damage that coffee growing there ceased to be economic. Since then, coffee rust has spread throughout the world, becoming established in South America during the 1970s (154).

Coffee rust resistance in existing *arabica* cultivars of coffee is highly race-specific, and therefore vulnerable to the emergence of new races of the fungus. *C. canephora*, *C. liberica* (a species that is both wild and cultivated to a limited extent) and other coffee species show better resistance and are being used in breeding programmes. In Colombia wild forms of *C. arabica* and of *C. liberica* are being used to confer resistance, but commercial cultivars have not yet been obtained (154a). Wild forms of *C. arabica* are also being used in East Africa (52b).

Rubber (22%)

Rubber (*Hevea brasiliensis*) is yet another crop with an extremely narrow genetic base. As far as is known, virtually all the trees grown outside the Americas are descended from a single collection of seeds made by Henry Wickham in the Tapajos area of Brazil in 1876. *H. brasiliensis* is the only *Hevea* species growing in that area, and had Wickham collected elsewhere in the Amazon basin he might have chosen the seeds of any of the other eight wild rubber species, which are less productive and less suitable for domestication.

Unfortunately, the Tapajos populations of *H. brasiliensis* have a lower yield potential than populations in the upper Amazon area, and since this was not discovered until the 1940s it probably delayed the development of high-yield trees. Equally unfortunately, the Tapajos populations are highly susceptible to South American leaf blight (SALB), caused by the fungus *Mycrocyclus ulei*. This disease has thwarted the development of the rubber industry in the Americas, and is the reason why less than 1% of the world's rubber is grown in the Western Hemisphere. It was a miracle that the *Mycrocyclus* fungus was not carried out of Brazil along with Wickham's seeds; if it had been, it is doubtful whether Southeast Asia's plantations could have been established (94).

Despite the narrowness of rubber's genetic base, breeders in Southeast Asia have achieved remarkable yield increases simply by selecting and crossing the best clones. Potential yield has risen from the 225 kilograms per hectare (196 lbs per acre) per year of Wickham's unselected seedlings to the 3,000 kg/ha/y (2,672 lbs/acre/y) of cultivars introduced in the 1970s (94). Additional genetic material, this time from the wild, will probably be necessary to achieve further yield increases, and will certainly be necessary should SALB ever invade Asia or Africa (from where so far it has been successfully excluded).

Since rubber is a perennial crop with a replacement cycle of as long as 30 years, it takes time for new germplasm to have a commercial impact. The most extensive use of wild species has been made in the Americas, in attempts to maintain profitable plantations constantly threatened by SALB. In Asia, three clones have been developed in Malaysia with wild germplasm, of which two have been released. Both have a measure of SALB resistance from wild *H. brasiliensis* (187a). Elsewhere in Asia (in Indonesia and Sri Lanka, for example), several wild species (*H. brasiliensis, H. benthamiana, H. pauciflora, H. spruceana*) are being used in breeding work, but so far with no commercial result (3a; 132a).

Tea (10%)

As far as we know, no wild germplasm has been used in tea improvement.

Cacao (9%)

Theobroma cacao, the plant that provides the world with cocoa and chocolate, demonstrates the potential of wild genetic resources for the improvement of commodity crops. Yields have been boosted through the incorporation of wild and semi-wild material from the upper Amazon basin into cultivars in West Africa, South America and Southeast Asia. In many cocoa-producing countries, these cultivars are the only officially recognised planting material. The wild germplasm supplies high-yield potential, ease of establishment, precocity (bearing fruit at younger age) and resistance to (or tolerance of) several diseases (137).

Fibre crops (21 million tonnes per year)
Cotton is the only fibre crop that has been improved with wild genetic resources.

Cotton (68%)

Four domesticated species produce the world's cotton. "Short staple" cotton (this refers to the length of the fibre) comes from *Gossypium arboreum* and *G. herbaceum*, and accounts for less than 1% of world production. Medium to long staple cotton comes from *G. hirsutum* and makes up almost 95% of world cotton supply. Extra long staple cotton comes from *G. barbadense* and accounts for the remaining 5%, most of it being produced in Egypt, the Sudan and the USSR (133).

G. hirsutum and *barbadense* have both been given valuable pest and disease resistance from their wild relatives. Virtually all cultivars of *G. barbadense* in the Sudan are resistant to bacterial blight (*Xanthomonas malvacearum*) thanks to genes from the wild cotton *G. anomalum* (GP3) (184a). In the Soviet Union, the Tashkent 1, Tashkent 2 and Tashkent 3 cultivars of *G. hirsutum*, occupying a million hectares (2.47 million acres), get their resistance to *Verticillium* wilt from the wild form of the species *G. hirsutum mexicanum* (GP1) (108).

Many insect pests are attracted to the cotton plant by "extrafloral nectaries" on the leaves and just outside the flowers. If these nectaries are absent, there are fewer pests; there is also less of the boll rot disease, which affects the bolls (fibre-filled seedpods) and can enter through the extrafloral nectaries (26; 113; 114). In the US, a no-nectary character from *G. tomentosum* (GP2) has been used in the Stoneville 825 cultivar of *G. hirsutum* to confer a degree of resistance to boll rot and many pests. This does not eliminate the need for pesticides, but it does mean that doses can be later, fewer and smaller (121).

Other fibre crops

We believe that there has been no successful use of wild genetic material to improve any of the other major fibre crops. With some of them, sisal (2%) for example, there is little current breeding work. With others, such as jute (21%, including jute-like fibres such as kenaf and roselle), there appears to be sufficient scope for improvement within the domesticated species themselves. Attempts have been made to transfer useful characteristics of wild species to flax (3%), but with no commercial result.

Timber

The application of genetics to silviculture (tree-growing) is quite recent: the intensive improvement of timber trees did not begin until after the Second World War (187). In many cases, candidates for plantations were selected from the wild without any thought as to the suitability of the trees for the environment in which they were to be grown. Sometimes the foresters were lucky. The available seed of *Pinus caribaea* and *P. kesiya* (two of the three most widely planted tropical pines in the world) just happened to be from populations (from Belize and the Philippines respectively) that were adapted to a wide range of environments and were productive under different types of management (22).

At other times, plantations of "improved" varieties failed or yields were well below expectations, because of the unsuitability of the planting material for the site.
* Mexican samples of *Pinus oocarpa* proved to be poorly adapted to many areas where they were introduced (22).
* A Himalayan population of *P. roxburghii* planted in South Africa had defective wood because of its spiral grain (182).
* The source material of *Eucalyptus microtheca* introduced into the Sudan produced "atrociously crooked stems" (182).
* Teak (*Tectona grandis*) planted in West Africa has been discovered to be inferior (104) as well as being drawn from genetically limited seed sources (80).

There are two distinct types of tree improvement. In the first selection, different *provenances* (geographical sources of germplasm; locally adapted populations) of a wild species are evaluated under particular conditions, and seeds, cuttings, grafts etc are taken from superior sources, being selected for seed production and planting. In the second, superior trees selected from within a species are crossed to enhance preferred characters such as rapid growth and good form.

Trees, of course, take longer to produce seed than most agricultural plants, and longer (10-80+ years) to reach harvestable age. So it can take a long time for tree growers to see the benefits of a breeding programme. Even for a fast growing tropical species such as *Pinus caribaea*, breeding can take many years. First, there is exploration and collection of seed from wild populations: 3-5 years. Then there is the evaluation of different samples in provenance trials: 5-15 years. Then superior trees must be selected for breeding: another 5-10 years. Even with overlap of these operations, the process requires a minimum of 20-25 years (182).

Foresters are patient people, and the results are already impressive. Coordinated international provenance trials of a number of tropical and subtropical species are being carried out as part of the Global Programme of Improved Use of Forest Genetic Resources of the UN Food and Agriculture Organization (FAO). Exploration and collection have been completed for 20 priority species: for three *Araucaria* species (parana pine, hoop pine, klinkii pine); for *Cedrela odorata* (Spanish cedar); for two *Eucalyptus* species (red river gum and coolabah); for *Gmelina arborea*; for *Tectona grandis* (teak); for *Abies cephalonica* (a fir); and for 11 *Pinus* species (the true pines).

The earliest of these trials began in the mid-1960s, with *Eucalyptus camaldulensis* (red river gum), the most widespread *Eucalyptus* species in Australia (30). Provenances grown in 21 countries have shown a ninefold difference in volume growth. Three in particular have demonstrated their superiority. One from Lake Albacutya (Victoria) has shown itself to be the best adapted to Mediterranean (winter rainfall) countries, increasing its volume by a third in just three years on one site in Morocco. A second from Katherine (Northern Territory) and a third from Petford (Queensland) grew outstandingly well in summer rainfall areas, and showed they were the best performers for Africa south of the Sahara (132; 183).

Trials with other species, although still in the early stages, have shown similar differences in performance. Some sources of *Pinus oocarpa* from Belize and Nicaragua are consistently among the fastest in early height growth (22). The progeny of a crossing programme of *P. patula* in Zimbabwe produced volume increases of 17% at five years and 37% at eight years compared with the standard yields of unimproved commercial trees (54). Seed source studies of white spruce (*Picea glauca*) in Canada have led to the identification of many rapid-growing races. Trees from the randomly selected seed of these races grow 15-20% taller than trees from local seed; if the tallest 10% of each race is selected as the seed source, then an additional 10% increase in height can be achieved (169).

Such benefits of the use of wild genetic resources easily outweigh the costs. For example, the cost of producing genetically superior white spruce seed in Canada is more than offset by 2-5% increases in yield of merchantable timber;

a 9% increase in yield gives a return of $5 for every $1 spent, and a 15% increase in yield can give a return of $36 a year per $1 a year invested (24). Among pines grown in southeastern US, improvement programmes have produced faster growth (reducing a 25 year rotation by two years), rust resistance, better tree form (especially straightness) and higher wood specific gravity, increasing the dollar-per-hectare value by an estimated 32% (4).

Forage crops

Alfalfa, *Medicago sativa*, was domesticated more than 2,000 years ago. But with a very few exceptions, the domestication of forage crops (grown for livestock to graze, rather than for humans to eat) began in the 19th century. Even today, much of the world's livestock is raised on natural or semi-natural, self-regenerated pasture. Domesticated grass and other pasture species are assuming an increasing importance, however, since their use can greatly increase the productivity of livestock production.

Along with this growth has come a greater emphasis on the use of the genetic variation within wild forage species, both as a means of improving crops already being grown, and as sources of new crops altogether. We have chosen the Australian experience to illustrate the actual and potential benefits of wild genetic resources for forage. Australia is the only major livestock-raising country in the world that grows forage crops in temperate, subtropical and tropical environments. Its experience with selecting and breeding forage crops is therefore unusual, but it is also indicative of the global situation.

Most of the 230 registered forage plant cultivars in Australia (9; 115) are selections of introduced species that have become naturalised, but 40% of them are selections from wild populations or have been bred using wild germplasm. From the point of view of pasture production, Australia can be divided into the tropical and subtropical north and the temperate and Mediterranean south. The north tends to get its species from tropical Africa and America; the south from Europe, North Africa and West Asia.

Livestock production on native rangeland is the predominant land use in northern Australia. The nutritive value of the native grasses fluctuates sharply over the year, peaking in the summer and slumping from autumn to late spring. During this long period of low feed-value, cattle suffer substantial losses in weight, and since the late 19th century attempts have been made to introduce forage species that would raise the productivity of the range. At first, the main emphasis was on grasses, but although productive species such as Rhodes grass (*Chloris gayana*) and guinea grass (*Panicum maximum*) did well in fertile areas, most of tropical Australia's soils are too poor to support them. The grasses currently used to improve pastures still have problems of quality and digestibility. The emphasis has now changed, though, to indirect improvement using legumes — the family of peas, beans, clovers and vetches

— both as highly nutritional forage crops in their own right, and as a means of raising the protein level and digestibility of the grasses by increasing the level of nitrogen in the soil (80a).

The results of adding legumes to grass are spectacular. When native pasture in Queensland was oversown with Townsville stylo (*Stylosanthes humilis*), the live weight gain of the cattle grazing it increased from 25 to 150 kilograms per hectare (22 to 134 lbs per acre) per year. Lucerne (=alfalfa, *Medicago sativa*) added to Rhodes grass (*Chloris gayana*) raised live weight gain from 90-179 kg/ha/y (80-159 lbs/acre/y). In other parts of the tropics, live weight gains of more than 700 kg/ha/y (623 lbs/acre/y) without irrigation and more than 900 kg/ha/y (803 lbs/acre/y) with irrigation have been reported (171). Such productivity is the direct result of using wild genetic resources.

As much as 56% of the certified grass seed and 90% of the certified legume seed produced in Queensland is of cultivars developed from wild selections. The top tropical grass cultivar, accounting for 41% of grass seed production (1978-80), is Basilisk, selected from wild signal grass (*Brachiaria decumbens*) from Uganda. The main legume species currently being sown are stylo (*Stylosanthes guianensis*) (46%), Caribbean stylo (*S. hamata*) (23%), and *Macroptilium atropurpureus* (19%). The four stylo cultivars and the single cultivar of *S. hamata* are selections from the wild. Siratro, the cultivar of *M. atropurpureus*, is the first tropical legume variety to have been bred in Australia rather than selected: the parental lines were wild plants collected in Mexico.

The most important resource available to plant breeders working with tropical forage crops is the enormous range of variation within the wild species. This enables the development of combinations of cultivars adapted to a wide diversity of growing conditions, and applies equally to established crops like stylo and to up-and-coming ones such as centro (*Centrosema pubescens*). The stylo cultivar Cook (obtained from wild material from Colombia) flowers and seeds much earlier than cultivar Schofield (derived from Brazilian material), and so can extend the area in which perennial stylo can be grown. Centro is at present restricted to the wet coastal fringe of tropical and subtropical Queensland, since varieties adapted to drier or colder districts are unavailable. Recently the CSIRO (Commonwealth Scientific and Industrial Research Organisation) Division of Tropical Crops and Pastures collected wild *C. pubescens* on the edge of its distribution. In the arid Guajira peninsula of Colombia, the plant collectors found a strain with a robust root system which enables the plants to survive and put on growth during mild droughts. And in Bolivia and southeast Brazil they found strains with a capacity for growth during cool seasons and an ability to withstand frosting to -3°C (26.6°F), characteristics that could extend the range of the crop into the subtropics (31).

Pasture sowing on a large scale is a modern development (since the 1950s), and the potential of wild genetic resources as sources of new forage crops and improvers of established ones is far from being fully realised. Dr EF Henzell, chief of CSIRO's Tropical Crops and Pastures Division, believes that the wild tropical forage species offer "great scope for selection and for later adaptation through breeding . . . [There] is so much valuable material still to be found and tested that the tailing off of new discoveries — the 'law of diminishing returns' — is not likely to operate in this field for at least another 20 years" (80a). Even then, the diminution of returns, when it ultimately occurs, will be in the rate of new crops or new cultivars developed per number of introductions (currently running at the rate of one for every 500). Judging from the experience of other crops, the impact of wild genetic resources, in terms of area grown to the improved cultivars and percentage increase in production, will continue to grow for long afterwards. According to Dr L. 't Mannetje, also of CSIRO's Tropical Crops and Pastures Division, most of the world's one billion plus (1,000 million +) hectares (2,500 million + acres) of tropical and subtropical grazing lands are unsuitable for other forms of food production, and less than 5% of it is at present fully productive. The total beef production of 7.7 million tonnes a year "could be doubled if only 25% of the area were improved through the use of legumes and appropriate fertilizer" (171).

In southern Australia, what Dr 't Mannetje calls the "use of legumes and appropriate fertilizer" has caused a revolution in farming dubbed the "sub and super" revolution after its two principal ingredients: subclover and superphosphate fertilizer. Subclover *Trifolium subterraneum*, a native of the Mediterranean region and of Europe north to England, was introduced to southern Australia several times during the 19th century. Overgrazing of the native rangeland together with use of phosphates favoured this and other introduced annual species at the expense of the native perennial grasses enabling Mediterranean annuals to replace Australian perennials (171).

T. subterraneum is the most important legume on acid soils in the south, growing on some 20 million hectares (49 million acres). Its counterparts on alkaline soils are the annual medics (*Medicago* spp). They are grown either in permanent pastures or in rotation with cereals; in either case, once sown or having spread from a nearby area they are self-generating. Their impact on the region's productivity has been remarkable. A 20-year study has shown that unfertilized land with little or no clover supported 2.5 sheep per hectare (one sheep per acre). Land top-dressed with superphosphate and growing vigorous subclover supported 13 sheep per hectare (5.2 sheep per acre). An equivalent study of the effect of the annual medics has not been made, but in three counties in South Australia wool production tripled in the 40 years since they were first sown (33; 33a).

All but two of the 19 subclover cultivars come from the naturalised populations. The two exceptions, cultivars Trikkala and Larisa, are of subspecies *yanninicum*, whose native area is limited to parts of Spain, Yugoslavia and northern Greece. Its value to Australian pasture production is far greater than its restricted natural distribution might suggest, because it has extended the use of annual clovers into areas with heavy soils and poor drainage.

Larisa is a selection of *yanninicum* collected from the wild in northern Greece. Among its more valuable qualities are substantial, though not total, resistance to clover scorch disease (*Kabiatella caulivora*). It also contains very low concentrations of the plant oestrogen formononetin. This quality is particularly important, because oestrogenic compounds in subclover and medics can reduce the fertility of ewes from 80% to 8% in a few years. Both qualities have been passed on to Trikkala, a bred cultivar with Larisa as one of the parents, described as "easily the most successful of the recently released subterranean clovers" (33a; 9; 115).

On Australia's alkaline soils the most important legumes are barrel medic (*Medicago truncatula*), strand medic (*M. littoralis*), disc medic (*M. tornata*), and gama medic (*M. rugosa*). Together they account for 26% of Australian production (1974-78) of Mediterranean annual legumes, compared with the 73% accounted for by subclover, a fair reflection of the relative importance of acid and alkaline soils in livestock production. While the naturalised populations have been the main source of subclover cultivars, virtually all of the annual medic cultivars have come from wild plants collected in their Mediterranean homeland. Although there are also naturalised populations of annual medics, most are unsuitable for commercialisation as they usually produce spiny pods that stick to the sheep's wool, costing an estimated $8 million a year.

Barrel medic is the most important of the four annual medics — accounting for 19% of Mediterranean Australia's annual legume production. Six of the 10 registered cultivars of barrel medic are collections from the wild (from Cyprus, Israel, Italy, Jordan and Tunisia), and one is a bred cultivar using wild parents (the rest are from naturalised populations). Each of the main wild genotypes have characteristics of special value.

* Cyprus: very early maturity, pods without spines.
* Israel: productivity and adaptation to marginal conditions.
* Italy: tolerance of spotted and blue green aphids.
* Jordan: high winter production and excellent recovery from harvesting.
* Tunisia: capacity to grow on heavier soils.

The flowering time of the genotypes also varies — from 66 to 136 days after sowing, the time being correlated with the length of the wet season at the collection site (the shorter the wet season the earlier the flowering) (33a; 9; 115).

Gama medic (only 1% at present) is becoming more important, because it is often more productive than other medics, and because many genotypes are tolerant of several insects. The three cultivars have all been developed from selections from the wild, two from Portugal, one from Italy. Paraponto from southeast Italy is valuable for its early maturity, good regeneration and large seeds. Sapo from Portugal has particularly useful tolerance of several pests: the blue green aphid (*Acyrthosiphon kondoi*), the spotted alfalfa aphid (*Therioaphis trifolii*), the lucerne flea (*Sminthurus viridis*), the red-legged earth mite (*Halotydeus destructor*), and the sitona weevil (*Sitona humeralis*).

Genes for tolerance of sitona weevil are very rare in the annual medics. In an examination of 850 genotypes of *M. littoralis* and 500 of *M. tornata* none was tolerant; only 14 genotypes (less than 1%) out of 2,500 genotypes of *M. truncatula* were tolerant. Among 19 annual *Medicago* species tested, *M. scutellata* had the highest proportion of tolerant genotypes: a mere 8.7% (13 out of 150). *M. rugosa* was the second highest with 7.4% (six out of 81 genotypes examined). Tolerance of aphids and sitona weevil is now the primary requirement for new annual medic cultivars, just as low oestrogen content is the primary requirement for new cultivars of subclover (9; 33; 115; 33a).

Wild germplasm is being used to improve feed as well as forage crops (crops grown to be fed to animals, rather than for livestock to graze). Three cultivars of lupin (*Lupinus angustifolius*), supplying 50% of Australia's lupin crop and worth US$20 million a year, carry a gene for resistance to the disease anthracnose (*Glomerella cingulata*) from a wild Mediterranean form of the species (GP1) (61b). The total contribution of forage and feed legumes to Australian agriculture far surpasses the cash value of any hay and grass feed, or of the livestock raised on all pasture. The structure and fertility of the soil is also improved. The value of the atmospheric nitrogen fixed by annual legumes in Australia has been calculated to exceed US$900 million a year (33b). With wild genetic resources supplying the raw material for 90% of the forage legumes grown in northern Australia, and almost 30% of those grown in southern Australia, it is clear that their contribution is immense.

Livestock
In contrast to crops, virtually no use has been made of wild genetic resources in livestock breeding. Attempts have been made to transfer adaptation to high altitude and genes for size and vigour from wild to domesticated sheep but with no commercial success (59).

Mammals and birds

In Israel, the Sinai desert goat (*Capra aegagrus*) has been crossed with the wild ibex (*C. ibex*). The resulting hybrid, called the ya-ez (combining the Hebrew words for the two animals), joins the former's hardiness and ability to go for days without water with the tastiness of the latter's meat (177).

In North America, the bison (*Bison bison*) has been crossed with cattle (*Bos taurus*) to produce a hybrid known as "beefalo" or "cattalo". It has been claimed that it survives better than cattle in exposed conditions, has a higher resistance to disease, calves more easily, converts food more efficiently, and produces leaner meat. The problem with this prodigal beast is that the males are more or less sterile. Half-bison bulls are invariably sterile; three-quarter and quarter breeds are usually sterile. To stand a reasonable chance of being fertile, a bull should have less than 25% bison ancestry.

The beefalo's sterility arises in a curious manner. One of the ways in which the bison is adapted to harsh winters is that its bull has a very small scrotum, and in cold weather it retracts its testes into its body. When the weather improves and the breeding season approaches, the testes are lowered into the scrotum once again. Apparently, hybrid bulls inherit the small scrotum but not the ability to retract the testes. Having a scrotum much smaller than that of domesticated cattle, the testes are held very close to the body and so are maintained at a higher temperature than in domesticated cattle throughout the breeding season. The heat may be sufficient to kill sperm, so rendering the bulls sterile.

Such problems may very well be overcome. The removal of undesirable traits that have been transferred along with the characteristics the breeder wants is standard procedure in crop improvement. Eventually we may see cattle with bison genes, along with sheep and goats and perhaps other animals improved with wild germplasm, but that time seems far off. At present, the world is estimated to have 3,778,000 farm mammals: mainly cattle (32%), sheep (29%), pigs (20%) and goats (12%), but buffaloes, horses and donkeys are also of major importance. Their improvement, together with that of 6,402 million farm birds (96% chickens, with ducks and turkeys also important) has been achieved entirely without the use of wild genetic resources.

Why should this be so? Sir Otto Frankel and Dr Michael Soule (59) have suggested six reasons:

* 1. Plants are earthbound, and can therefore be expected to have more specific (and hence more easily identified and transferred) adaptations to particular environments than mobile animals.
* 2. Many plants can be self-fertilized but all livestock species are heterosexual.
* 3. The evolutionary distance between wild and domesticated species is less in animals than in plants, and hence there is greater scope with the animals for selection and breeding within the species and less need to return to the wild.
* 4. Domesticated plants and animals differ greatly in their response to diseases and parasites and the means available for their control.
* 5. There is a greater abundance of wild relatives of plants than of animals, many of the progenitors and other close relatives of domesticated animals being extinct or rare.
* 6. Use and maintenance of the genetic resources of wild animals pose much greater logistical problems than do those of wild plants.

The first three of these six reasons seem less persuasive than the last three.

On the first reason, many traditional breeds of cattle and sheep, for example, do manifest adaptations to their local environments (19; 97; 106), as do their wild relatives.

On the second reason, while many of the crop species that have been improved with wild germplasm are indeed inbreeders (they can be self-fertilized, such as oats, wheat, bell pepper, tomato, tobacco), at least as many are outbreeders (they have to be fertilized by other plants, as is the case with sugar beet, sunflower, sugarcane, oil palm, potato, cacao) (137).

To substantiate their third reason, as examples of evolutionary distance between wild and domesticated plant species, Frankel and Soule (59) point to the solid rachis (spindle) of the ear of wheat or the single large cob of maize, which would prevent their survival in nature, because seed dispersal depends on human agency. But similar differences can be found in some livestock species. For example, the turkey has been bred (using artificial insemination) so successfully for a large amount of breast meat that males are prevented by the size of their breast from mounting females, and so are unable to mate (131). In plant improvement, evolutionary distance seems to be a factor in determining whether selection rather than breeding proper is resorted to; but once breeding is the established procedure, evolutionary distance does not seem to influence the choice between domesticated or wild genetic resources. The former are invariably preferred to the latter, and the wild genetic resources are used only when they appear to be the sole source of whatever characteristics are sought.

By contrast, the mechanisms of disease and parasite control — which is probably the most important type of characteristic supplied by the wild relatives of crops — do indeed differ markedly between plants and animals, as Frankel and Soule argue in their fourth reason. Animals are able to develop immunity to a disease, and to retain it for some time, while remaining genetically susceptible; immunity can also be induced by inoculation. Plants do not have this capacity, but they often achieve disease resistance by jettisoning affected cells or other parts (a mechanism called hypersensitivity), something that animals cannot do. Plant cultivars are genetically much more uniform than animal breeds and so are more vulnerable to virulent strains of a disease. This last difference may diminish, however, with the trend toward greater uniformity among and within livestock breeds (59).

The problem of the supply of wild livestock genetic resources is also very real, as the fifth reason argues. As we shall see in Chapter 5, many of the closest wild relatives of domesticated animals are extinct, endangered or rare. For example, of the five species of the cattle genus *Bos*, the wild progenitor of cattle (*B. primigenius*) is extinct; the kouprey (*B. sauveli*) and the wild form of the yak (*B. mutus*) are endangered; and the banteng (*B. javanicus*) and the gaur (*B. gaurus*), wild forms of the Bali cattle and mithan respectively, are vulnerable (= threatened in the US) (36; 170).

The maintenance and use of wild livestock genetic resources also pose far greater logistical problems than exist with plants. A plant breeder can readily assemble a wide array of different forms of the crop and its wild relatives. Some crops are certainly difficult to work with, such as vegetatively propagated plants (sugarcane, for example) or tree crops that can only be maintained as trees rather than as seeds (cacao, for example). But even these present fewer and lesser difficulties than most animals, for which safe routine procedures for preserving either semen and ova (sperm and eggs) or embryos have yet to be developed. So the breeding pool for many livestock species has to be kept in the form of live animals, which greatly reduces the number of individuals the breeder can work with. Similarly, cost and often also the risk of spreading infection restrict or entirely prevent the selection procedures in animal breeding that have been so successful with plants. Quarantine restrictions limit the movement, and hence the use, of animals to a much greater extent than they do of plants (59).

Insects

The two insect livestock groups of major world importance are honey bees and silkworms. Honey bees (*Apis mellifera*) produce about 1,000,000 tonnes of honey each year. Although they have been domesticated for at least 5,000 years, genetically they are still very close to the wild. There are many strains of honey bee, and there is great variation within the strains for characteristics such as honey-production, wintering ability and disease resistance. But efforts to improve bees, using the selection techniques that have been successful in mammal, bird and plant breeding, resulted until recently only in weak, unproductive colonies (156). Better understanding of honey bee genetics has led to some improvement, but the use of wild bee germplasm seems a long way off.

World production of silk is more than 60,000 tonnes a year. Most is produced by the mulberry silkworm (*Bombyx mori*), a species that has been domesticated almost as long as the honey bee. The other main silkworms are either completely wild (the *Anaphe* silkworms) or only partially domesticated: the eri silkworm (*Attacus ricini* or *Philosamia ricini*) and the tasar or tussah silkworms (*Antheraea* species). *Anaphe* silk production is not included in world statistics, and in parts of West Africa where it occurs the silkworms are probably more important as food than as silk producers (178). The world figures include eri and tasar silk but do not identify the different sources of silk. Tasar silk production has been estimated to account for 4,000 tonnes a year, or almost 7% of the total (99).

A great deal of breeding work has been done with the mulberry silkworm, and there are a great many domesticated strains. No use has been made of wild germplasm, since wild forms of *Bombyx mori* are not known to exist (156). Tasar silkworms, however, are in the process of domestication and in India wild forms are being used in their improvement. *Antheraea roylei* from India and *A. pernyi* from China have been crossed to produce the hybrid *A. "proylei"*. Silk from the hybrid is claimed to be the finest ever produced in India and the new species also outyields its parents by almost 170% in weight of silk and by 94% in filament length. The hybrid is expected to play a major part in efforts to expand the tasar silk industry and increase its capacity to provide rural communities with income and jobs (98; 99).

Aquaculture

Although aquaculture is almost as old as agriculture, very few aquatic species have been truly domesticated. The carp (*Cyprinus carpio*) was domesticated by the Chinese sometime between 4,000 and 2,500 years ago; *Tilapia* species may have been raised by the Sumerian and other civilisations of West Asia even earlier. Oysters were cultured by the Romans 2,000 years ago, but this form of "culture", as with the raising of most aquatic animals and plants even today, amounted simply to enhancing the survival of wild-propagated organisms. It does not qualify as domestication, and many authorities regard only two aquatic species as completely domesticated: the carp and the rainbow trout (*Salmo gairdneri*). Many other species are in the process of domestication, but their stocks have been collected from the wild with little or no regard for the relationship between the genetic make-up of the source populations and their potential productivity, product quality, disease resistance and other characteristics. The genetic variation of aquatic animals and plants is a great and almost entirely untapped resource.

There are nevertheless several examples of the use in aquaculture of wild genetic resources, which suggests a considerable potential. In the USSR, the domesticated carp has been crossed with the wild Amur carp (*C. carpio haematopterus*), and the latter's cold resistance has enabled carp production to be extended as far north as latitude 60°N (8). In Hungary, the grass carp (*Ctenopharyngodon idellus*) and the big head (*Hypophthalmichthys nobilis*) have been crossed to produce a sterile hybrid; it can be introduced into lakes and rivers to control weeds with no danger of overpopulation or hybridisation with native species (47).

Overpopulation is a particularly serious problem with *Tilapia* species raised by aquaculture, because it results in the production of large quantities of very small fish. Among the possible ways of preventing overpopulation are separating parents and young (time-consuming and expensive), introducing predators to the ponds, and monosex culture. Monosex culture is the stocking of the pond with individuals of one sex. Males are preferred because they continue to grow during breeding periods (females do not) and also all-male crosses are easier to achieve than are all-female crosses. Males are not without their problems, though. Bardach, Ryther and McLarney report (8) that "whether or not there are females present, (the males) optimistically construct spawning nests. The favoured location for nest building is at the base of the pond bank, and the nest-building activities of a great number of males may eventually undermine the bank."

Hybrids consisting entirely of males can be achieved by crossing male *Tilapia macrochir* and female *T. nilotica* (difficult to do); male *T. hornorum* and female *T. mossambica* (*T. hornorum* is hard for many aquaculturalists to obtain); and female *T. mossambica* with male *T. nilotica*. An advantage of the *mossambica* x *nilotica* all-male hybrid is its hybrid vigour, but it is also unusual in that only one source of each of the parental species can get the desired results. To get 100% males, the male *mossambica* must be from Zanzibar and the female *nilotica* must be from Lake Albert. Other sources of either species are not effective (8).

More research is needed on the distribution of genetic variation within aquaculture species, but the possibilities can be illustrated by some of the wild trout species. The Kamloops strain of rainbow trout (*Salmo gairdneri*), native to inland British Columbia, Canada, outgrows all other strains (8). Steelhead (sea-growing rainbow) from certain streams in British Columbia and Alaska, and the Lahontan strain of cutthroat trout (*S. clarkii henshawi*) found only in two lakes in Nevada (US), can grow to very large sizes, possibly due to genetic factors (8).

Great differences have been found in growth rate, disease resistance and tolerance of increased acidity due to acid rain among and within strains of brown trout (*S. trutta*) in Norway (61). This last example shows that the wild genetic resources of aquatic species have potential not only for the development and improvement of domesticates but also for the maintenance and enhancement of stocks in the wild.

3. The nature of wild genetic resources

Certain patterns emerge from the last chapter that may serve as a guide to what we can expect from wild genetic resources in the future. They concern the rate of use of wild genes; the kinds of benefits they provide; and the kinds of wild species being used.

The use of wild genetic resources is clearly on the increase, as Figure 1 shows.

Two conclusions can reasonably be made from the table. First, the number of crops drawing on wild genetic resources is increasing. Second, the use of wild germplasm is seldom a one shot deal; over time, cultivars containing wild germplasm tend to account for an increasing proportion of the crop (as in sugarcane), or a growing number of wild species are used (as in potato), or both (as in tomato). As would be expected, wild genetic resources have on the

*Figure 1: the decade in which commercial cultivars improved with wild germplasm were released for the first time. Lower case letters (*wheat*), indicate cultivars with wild germplasm which account for less than 10% of current world production. Capital letters (*RICE*) indicate cultivars with wild germplasm which account for 10-50% of world production. And bold capital letters (**OIL PALM**) indicates cultivars with wild germplasm which account for more than 50% of current world production. An asterisk (**CACAO***) indicates a crop which has obtained more than one type of characteristic (such as disease resistance, high-yield potential, cold hardiness) from the wild. We have had to rely on a certain amount of informed guesswork in assigning crops to these categories.*

1870s:	**GRAPES***
1900s:	-
1910s:	**POTATO***, strawberry
1920s:	OATS, **SUGARCANE***
1930s:	**TOMATO***
1940s:	sugar beet
1950s:	**SUNFLOWER*, OIL PALM**, TOBACCO
1960s:	wheat, maize, barley, carrot, peas, **CACAO***, cotton
1970s:	RICE, cassava, sweet potato, bell pepper, apple, rubber, sesame

whole made their greatest contributions in those crops which have used them longest.

Another generalisation that can be made is that the more important the crop, the greater the breeding effort, and the greater the likelihood that genes from wild species will be used successfully. Three-quarters of the crops listed in the table (18 out of 24) account for more than 10% of world production of their particular crop category (eg cereals, oil crops), the exceptions being oats (3% of cereals), sesame (2% of oil crops), peas, carrots and bell pepper (4%, 3%, and 2% respectively of vegetables and pulses), and strawberry (1% of fruits and nuts).

By the same token, very few of the major crops have *not* been improved with wild germplasm: only soybean among the oil crops, cabbages and their allies among the vegetables and pulses, bananas and citrus fruits among the fruits, coffee and tea among the commodity crops, and jute among the fibre crops. For some of these, new cultivars using wild genetic resources are being developed (two, banana and coffee, are discussed in Chapter 2). In others, notably soybean, the genetic base is so narrow that wild relatives can be expected to make a contribution once breeding problems have been overcome.

The fact that a crop is "minor" from the lofty perspective of international statistics does not exclude its improvement with wild germplasm. Lettuce, hops, currants, gooseberries, raspberries and blackberries have all benefited greatly from breeding with wild genetic resources; and vanilla and black pepper are among the crops that can be expected to benefit in the near future. In addition, as breeders of timber, forage and aquatic resources change their emphasis from species to gene pools, use of wild genetic resources in those areas has begun and will almost certainly increase.

Benefits of wild genetic resources

By far the most important benefit obtained from the wild has been disease resistance. There are several reasons for this. Generally, resistance obtained from wild species is under the control of a single dominant gene. It is simply inherited and easily transferred. It is also clearly expressed: while a breeder may have to undertake careful measurements over several years to discover which plants show high yield, it is usually quickly obvious which plants suffer from a disease.

When breeders want a particular characteristic such as disease resistance, they look for it first in the locally available cultivars, because these already have all the other characteristics of yield, quality and environmental adaptation that are needed in the crop. If disease resistance is not available in the local cultivars, then the breeder will look elsewhere in the crop, among the cultivars of other countries. Even if these are not adapted to local growing

Figure 2: plant characteristics obtained from the wild, and the number of crops now having each characteristic.

Disease resistance	18
Pest resistance	5
High yield	5
Vigour	4
Environmental adaptations	4
High starch/soluble solids/vitamins	3
Cytoplasmic male sterility	1
Petaloid male sterility	1
Harvest and transport adaptations	1

conditions, at least they are likely to have the yield and quality factors that growers and consumers expect. Wild relatives of the crop will be used only as a last resort.

This reluctance to use wild species is only partly due to the difficulties of crossing — which range from non-existent in the case of the primary gene pool to almost insuperable in the case of the tertiary gene pool. But even if disease resistance proves to be available in a wild species, and even if this wild species is in the crop's primary gene pool and is therefore completely inter-fertile with it, the breeder can still have problems.

The main difficulty is what might be called the package-deal effect. Suppose you are a tomato breeder. You and your colleagues have developed an array of tomato varieties that yield phenomenally, taste superb, and drop obligingly and intact into the arms of mechanical harvesters. Unfortunately, 20% of your crop is lost to the fungus disease stinking smut every year. You discover a wild tomato that yields erratically, tastes disgusting, and is in almost all respects a grower's nightmare, but it *is* resistant to all known races of stinking smut.

The cultivars and the wild species cross easily, giving progeny that express the characteristics of the cultivars and of the wild species in varying combinations and degrees. Through a programme of selection and crossing among several generations, you expect to recover all the good characteristics of the cultivars and to incorporate the disease resistance of the wild species. Unfortunately, the genes for resistance are linked to genes for other undesirable characteristics of the wild species (perhaps the disgusting taste), and you find it extremely difficult, perhaps impossible, to separate them. This is the package-deal effect. You cannot get one lot of genes without the other.

This difficulty is much less likely to occur when the resistance is controlled by a single major gene than when it is controlled by several minor genes. In the former case, the second generation progeny will be easily separated into those that are resistant and those that are not. The resistant ones can then be backcrossed to the original cultivars, and their resistant progeny can again be backcrossed to the cultivars, until cultivars are achieved that are virtually identical to the originals except that they are disease resistant. But if the resistance is controlled by several genes, the second generation progeny may show every conceivable gradation between susceptibility and resistance. The selection and crossing process then becomes more complicated and protracted, quite apart from any package-deal effect.

Recovery of desirable characteristics can be even more difficult if the crop is an outbreeder, like maize or banana, which needs another line to fertilize it, rather than an inbreeder, like tomato or wheat, which fertilize themselves. Outbreeders may simply be self-incompatible (that is, they cannot be crossed with themselves, thus preventing backcrossing) or if self-compatible, they are likely to suffer inbreeding depression, leading to lower yields and vigour after a few generations.

This explains why most of the useful characteristics that have been obtained from wild species have been controlled by single genes. They include fertility restoration in sunflower; petaloid male sterility in carrot; the stalk modification in the tomato that adapts it to mechanical harvesting; the high-yield factor in the oil palm; and much of the disease and pest resistance in most crops. There are some important exceptions, however. The high-yield potential and vigour of sugarcane, cacao and potato, for example, are due to the interaction of whole genotypes rather than to specific genes. Some forms of resistance, such as the field resistance to late blight transferred to potatoes from *Solanum demissum* and the resistance to *Cercospora* leafspot transferred to sugar beet from *Beta maritima* are controlled by several minor genes.

Many fruit and nut crops, such as grapes, are sometimes easier to breed because they can consist of two cultivars, the scion (fruiting) cultivar and the rootstock cultivar. Resistance to diseases and pests that affect the lower part of the plant can be achieved by improving the rootstock without worrying too much about the effect on fruiting quality. Great grape cultivars like Cabernet Sauvignon, Pinot Noir, Chardonnay, and Reisling can continue to be grown, their quality uncompromised by alien genes, thanks to the pest resistance and other adaptations of their wild-derived rootstocks.

Most characteristics that breeders seek are available to them within the crop. That breeders have already turned to wild species to the significant extent they have, is a measure of the rarity and importance of the characteristics concerned. There is nothing special about wild species that makes their disease resistance "better" than the disease resistance obtainable from cultivated varieties. They simply represent an expanded supply of potentially useful and more or less accessible genes. The law of supply and demand applies to wild genetic resources: when demand for a genetic characteristic is high and the supply of that characteristic in the crop is low, then it is sought from the wild.

Pressure for a wild genetic resource can come from the demand side or it can come from the supply side. Disease resistance illustrates the first. Diseases have become such a problem in agriculture that resistance to them has become a major breeding objective. However, it is only one of several objectives. If resistance to a disease can be won only at the expense of other objectives — such as superior quality — then the growers and breeders may decide to live with lower levels of resistance.

Fungal diseases are an exception. They pose such an acute threat to certain crops that resistance to them has become the priority breeding objective, outweighing all others. Stem and leaf rusts of wheat in North America are examples. The fungi which cause these diseases have much greater genetic variability than pests or virus and bacterial diseases. New strains and races of fungal disease, unaffected by the resistant genes so laboriously introduced into the crop, cause serious and repeated problems. So the disease resistance existing within many crops (as in wheat, potatoes, cotton, sugarcane and others) is not sufficient, and wild sources are also needed. Disease resistance is therefore the most common contribution of wild genetic resources, and it is likely to be the first contribution in the future to crops that have yet to be improved with wild germplasm (such as peanuts, chickpeas, cucumbers and squashes).

There are broadly three circumstances in which pressure for a wild genetic resource can come from the supply side. One is the case of a recent but prescientific domesticate, such as oil palm or rubber. Both crops were domesticated in the last century. Their domestication was prescientific in the sense that no attempt was made to seek out the gene pool within the species that was best suited to domestication. That rubber and oil palm have done as well as they have, is sheer good luck; that they could have done much better was sheer bad luck. But sooner or later breeders of such crops have to return to the wild species to build a broader genetic base. For oil palm this has already occurred, but a similar return to the wild can be expected for rubber.

The second situation in which a crop itself may be unable to supply the genetic resources needed for its improvement is when its variability is reduced. This can happen when it goes through a bottleneck, as was the case with the potato and the tomato. It can also happen when an outbreeding species becomes inbred, as a result of using a limited number of parental lines over and over again. Sugarcane in the last century is an example.

Sometimes there are other domesticated forms on which crop breeders can draw. This is notably so with the potato, because outside South America the crop developed independently of the neo-tuberosum group of traditional South American cultivars. To a much lesser extent it was also true of sugarcane, which was able to draw on the *sinense* domesticated forms. Basically, however, the principal resource of the sugarcane or the tomato breeder is the much greater diversity of the wild species.

The third type of problem that conditions the supply of genetic resources is rarity. Some genes are rarer than others, and some genes are very rare indeed.

An example is the j_2 gene for the jointless pedicel (stalk) in the tomato, which allows mechanical harvesting: this has only been found in one single population of *Lycopersicon cheesmanii* in the Galapagos Islands (142). Similarly, only four sources of cytoplasmic male sterility have ever been found for sunflower, all in wild species (137). Such rare genes have a greater chance of occurring in the wild, because the wild gene pools are generally more numerous and diverse than the domesticated pools.

What kinds of wild species are used?

A simple rule seems to govern the use of wild species in crop improvement. The more closely related the wild species is to the crop, the more likely it is to be used.
* Of the 24 major world crops that have been improved with wild germplasm, the primary gene pool (GP1) has been the germplasm source for 18: rice, barley, oats, sweet potato, sunflower, oil palm, sesame, tomato, carrot, bell pepper, grapes, apple, strawberry, sugarcane, sugar beet, rubber, cacao, cotton.
* The secondary gene pool (GP2) has been the germplasm source for eight of the crops: maize, potato, cassava, tomato, peas, sugarcane, tobacco, cotton.
* The tertiary gene pool (GP3) has been the germplasm source for three of the crops: wheat, tomato, cotton. (The total adds up to 29 rather than 24 because breeders have turned to two of the gene pools of sugarcane and/or three gene pools of tomato and cotton.)

The popularity of the GP1 species is easily explained. Since they are completely inter-fertile with the crop, they pose less of a problem to the breeder than do the GP2 and GP3 species. Given the occurrence of a particular characteristic in the same degree in the GP1 species and GP2 species, the breeder will obviously choose the former over the latter. As a rule, the GP2 and GP3 species are poorly known, so breeders are unaware of what these gene pools could contribute. They are generally even less adequately represented in germplasm collections than are the GP1 species. Breeders may have difficulty getting hold of sufficient material, even if they are aware of its qualities, and even if they have the time and resources to overcome crossing problems.

The future of wild genetic resources

The trend towards wild genetic resources can be expected to continue and probably increase. As domesticated plants and animals become more dependent on the breeder, to help them cope with new ecological conditions,

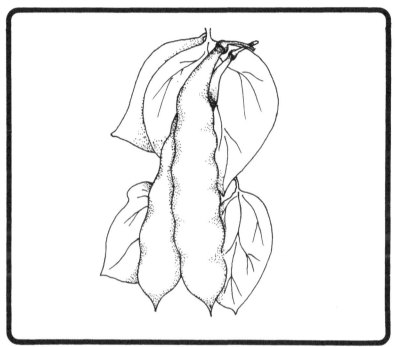

BEANS. Breeders have tried unsuccessfully to transfer the cold resistance and vigorous root systems found in wild forms of runner bean to the cultivated common bean Phaseolus vulgaris *(above), which gives us "baked beans".*

more virulent diseases, peskier pests, fluctuations in climate, and changing economic demands, so the breeders will go more often to the wild species for the rich store of genetic resources they contain.

It is likely that tomato breeders, for example, will continue to turn to the wild relatives, since most of the variation is there rather than in the cultivars. Increasing emphasis on wild sources of germplasm is also likely in those crops in which, even though they are quite diverse, the particular qualities required by the breeder are missing. The main source of pest and disease resistance in peanuts, for example, is now the wild species (139).

Crops like sunflower have already benefited substantially from wild germplasm; they now rely on the wild to protect them from the potentially damaging consequences of their own improvement. Because the hybrids that make up most of the sunflower crop share the same cytoplasm, a large part of the crop (90% in North America) could be destroyed at once if a disease emerged to which that cytoplasm was susceptible. This happened to hybrid maize in the US, which like sunflower is also produced by the cytoplasmic male sterility method. In 1970, maize yields in the United States were cut by an average of 15% and farmers lost hundreds of millions of dollars because of the susceptibility of the maize T cytoplasm to southern cornleaf blight. (*Helminthosporium maydis*). Fortunately, to guard against a similar event occurring in sunflower, alternative male sterile cytoplasms have already been found in wild sunflowers and are being bred into parental lines.

Changes in technology and in social expectations are other encouragements to use the greater diversity available in wild species. Cotton breeders are turning to wild cotton species for better fibre, because they now have the technology to evaluate lint quality precisely. And they look to wild cotton for pest and disease resistance, now that higher environmental standards and more sophisticated awareness of economic and ecological costs discourage reliance on pesticides (139).

The rate of uptake of wild genetic resources could be profoundly influenced by developments in prebreeding and genetic engineering. Prebreeding means the transfer of germplasm from wild species to parental lines of the domesticate. We have already commented on the obstacles to using GP2 and GP3 species. Overcoming such obstacles is often beyond the capabilities of many breeders. Prebreeding incorporates wild germplasm in plants that behave in ways with which breeders are familiar, so they can more readily use it.

Some prebreeding is already being done by the big international agricultural centres and by major national institutions.

* The Centro Internacional de la Papa (CIP: International Potato Centre) in Peru has a programme of wide crosses with the more distantly related wild potato species, followed by backcrossing with cultivars (76a).
* The Faculté des Sciences Agronomiques de l'Etat of the Université de Gembloux (Belgium), the world centre for *Phaseolus* beans, has a similar programme. The laboratory develops hybrids, studies the early generations, overcomes any problems of fertility restoration, and then sends the material to international centres for detailed evaluation (117c).

The prebreeder bridges the gap between cytogeneticists, pathologists and other specialists on one side and the cultivar breeder on the other, and delivers the desired characteristics obtained from the wild gift-wrapped in a domesticated genetic package. Prebreeding is probably the major contribution that public breeding programmes can make in general, and that international and developed country programmes can make to crop improvement in developing countries.

Transferring germplasm from GP3 wild species brings breeding very close to genetic engineering. One method involves the use of bridging species: incompatible wild species X is first crossed with compatible wild species Y (the bridging species) and the resulting hybrid is then crossed with the domesticated species Z. More radical methods involve doubling the chromosome number with colchicine, a chemical from the autumn crocus; zapping chromosomes with radiation so that they break and form new combinations; extracting the embryo from the seed and culturing it in the laboratory.

All these cases, however, still involve genes that are transferred sexually, so they are in the last resort not genetic engineering but breeding. The term genetic engineering implies "no sex please, we're engineers". Among the non-sexual techniques developed by the engineers are:
* protoplast fusion, in which the cell walls are removed from the cells, usually by enzymes; the resulting wall-less cells, or protoplasts, can then fuse with each other;
* using special vectors, such as the bacterium (*Agrobacterium tumefaciens*), to introduce alien genes into the plant's cells;
* culturing plant cells directly without bothering to grow them into plants.

The first two techniques are basically just additions to the current battery of radical methods for transferring genes from GP3 species. Converting the new germplasm sources into superior cultivars still remains the major problem. The third technique seems to be the most novel development. Crops that are grown for their chemical content (medicinals, essential oils, sweeteners, insecticides, and so on) could be cultured in the factory rather than grown in the field.

The following quote from a report on applied genetics by the Office of Technology Assessment of the US Congress (131) illustrates the potential:

"The vinca alkaloids — vincristine and vinblastine for instance — are major chemotherapeutic agents in the treatment of leukaemias and lymphomas. They are derived from the leaves of the Madagascar periwinkle (*Catharanthus roseus*). Over 2,000 kilograms of leaves are required for the production of every gram of vinca alkaloid at a cost of about \$250/g. Plant cells have recently been isolated from the periwinkle, immobilized, and placed in culture. This culture of cells not only continues to synthesize alkaloids at high rates, but even secretes the material directly into the culture medium instead of accumulating it within the cell, thus removing the need for extensive (= expensive?) extraction procedures."

Useful variations have also been found in different populations of *Rauvolfia*, the source of reserpine (used to treat hypertension) and other medicinal alkaloids. The range of total alkaloids in the African species *R. vomitoria* from Senegal to Uganda runs from 0.2% to 3.0%. The highest alkaloid content, as well as the largest proportion of reserpine (0.3%), was found in samples from Zaire. Of the two forms of the Asian species *R. serpentina* in India, one is high in reserpine but low in total alkaloids, and the other is the opposite (139).

Similarly, genetically controlled variation in alkaloid content occurs in the poppy species *Papaver bracteatum*, growing wild in northwestern Iran and adjacent areas of the USSR. *P. bracteatum* shows promise as a new source of codein that could replace the opium poppy *P. somniferum*. Codein is one of the most important alkaloids in medicine, used to relieve pain and suppress coughs. Its manufacture is the main justification for much of the legal production of opium, from which it is currently synthesised.

The advantage of *P. bracteatum* is that it produces neither opium nor morphine (the precursor of heroin), but the alkaloid thebaine. Thebaine can be converted to codein just as easily as can opium, but is extremely difficult to convert to opium or morphine. Hence substitution of *P. bracteatum* for *P. somniferum* as the source of the starting material for codein manufacture offers a means of reducing the illegal trade in opiates. Some gene pools of *P. bracteatum* have higher concentrations of thebaine than do others.

As petroleum prices rise and as better ways of extracting and using plant chemicals are developed, so plants can be expected to grow in importance as sources of fuel and chemical feedstocks. Biochemicals from plants (and possibly also from animals such as corals and sponges) may be produced conventionally or by cell culture. Among potential biochemical crops are latexes (guayule and several Euphorbia species) both for rubber and fuel, waxes (*Candelilla* and *Calathea lutea*), and oils (buffalogourd, crambe, jojoba, meadowfoam, stokes aster, vernonia, and several others). The successful domestication of these species will depend heavily on the availability of suitable germplasm.

Identifying the most useful gene pools within the species to be domesticated is now recognised as being as important as deciding which species should be domesticated. Part of the promise of jojoba (*Simmondsia chinensis*) and meadowfoam (*Limnanthes alba*) lies in the availability of considerable diversity within the species (and also, in the case of meadowfoam, in related species). Some wild stands of jojoba appear to be unusually productive and these are being used in selection programmes. Similarly, populations of meadowfoam show marked variation in fatty acid content and in agronomic characteristics such as growth habit and seed retention. Conversely, lack of germplasm has impeded domestication of another potential oil crop, crambe (*Crambe abyssinica*) (137).

The potential of wild genetic resources is therefore in three main areas: in the culture of biochemicals, in the development of new domesticates, and in the improvement of existing domesticates. Because we have concentrated on crops and livestock of major world importance, we may have given the impression that the last of these — improvement of domesticates — involves only a small number of species. This is so only with respect to livestock. Ten mammal and bird species together account for 98% or more of world livestock production, and an additional 30-40 species are regionally significant (eg, camel, llama, reindeer, mink, guineafowl). By contrast, at least 70 plant species are of major world importance as crops (excluding timber trees) and there are hundreds more that are regionally important. The wild species that could be used in their improvement number in the thousands. The contribution so far of wild genetic resources to existing domesticates is just the tip of the iceberg.

Similarly, new and potential livestock domesticates are few (eg, red deer in New Zealand, musk deer in China), but there are some 100 candidate aquaculture species and several hundred incipient plant domesticates. In the tropics alone more than 250 pasture grass and legume species are in the course of domestication, and there are at least 120 species of timber trees with substantial potential for tropical and subtropical plantations. Preliminary results of a study by us of domestications suggest that current rates of successful plant domestication are historically high, reflecting the new demands of a rapidly changing society and more scientific use of wild genetic resources (137).

The final factor that will help determine the future of wild genetic resources is their conservation. Breeders and genetic engineers can devise more and more ingenious ways of using available genes but they cannot create new ones. Although wild genes are a brand new resource and although their utility is growing dramatically, the lack of progress in conserving them (Chapter 6) casts a shadow over their future.

4. Where are wild genes found? And who uses them?

The crops and livestock produced in one country may rely on the domesticated genetic resources of another country, and on the wild genetic resources in a third. So the costs and benefits of using and maintaining genetic resources, wild or domesticated, are unequally distributed. Some countries may find themselves giving more than they are receiving, and vice versa. This problem is compounded by questions of ownership, particularly the current trend towards private ownership of cultivars and breeds and hence, it is feared, of the genetic resources they contain.

The two issues — international inequalities in the distribution and use of wild genetic resources; and private ownership of the fruits of using wild genetic resources — are distinct, and we will discuss them separately.

Who has got them? And who benefits?

The view is increasingly expressed that on the whole wild genetic resources are found mostly in developing countries, which must pay for conserving them without much reward. Certainly, some of the crops that benefit from wild genetic resources are grown in developing countries, but many are grown in the developed countries. So it is argued, the developed countries do very nicely out of wild genetic resources, without having to sacrifice scarce land for their conservation.

This view has been succinctly stated by Indonesia's population affairs minister Dr Emil Salim when he was minister for development supervision and the environment: "In a sense, we in the South are conserving our genetic resources for the North to exploit and enjoy."

This thesis has been elaborated by Pat Mooney in *Seeds of the earth* (123), a critique of the status of international maintenance, exchange and use of genetic resources. "Virtually everything people eat", he writes, "can be traced back to fewer than a dozen centres of extreme genetic diversity — the so-called 'Vavilov Centres', named after the great Russian scientist who dominated botany in the 1920s". Vavilov's eight centres of crop origins (China; India + Burma + Southeast Asia; South-Central Asia; West Asia; Mediterranean; Ethiopia; Middle America; and South America) lie almost entirely in developing countries. The exceptions are the northern parts of the South-Central Asian and West Asian centres (called by Vavilov the Central Asian and Near Eastern centres) which are in the USSR; and the northern half of the Mediterranean centre which is in southern Europe.

This view is oversimplified. With growth in knowledge since Vavilov's time, other scientists have expanded his centres, culminating in the scheme of PM Zhukovsky: 12 "megagene" centres covering all of the US, almost all of Europe and Africa as well as virtually all of the areas included by Vavilov (188). JR Harlan's revision reduced Vavilov's list to three centres (West Asia; China; Mexico) plus three "noncentres" (middle Africa; South and Southeast Asia; South America), these being ancillary areas too large to be called centres (66).

A second difficulty is one that Vavilov himself recognised: centres of *origin* are not necessarily the same as centres of genetic *diversity*. Ethiopia, for example, is certainly a centre of diversity for wheat, barley, peas, lentils and flax, but none of these crops is believed to have originated there (191). Bread wheat (*Triticum aestivum*) is believed to have originated in West Asia (the trans-Caucasus), where there is a centre of diversity, but there are other centres of diversity in Japan, Korea, China, India, Afghanistan, North Africa and Europe (188).

However, the real difficulty about wild genetic resources is that centres of genetic diversity of crops, and centres of genetic diversity of their wild relatives, do not necessarily coincide. In fact, "centres" may not exist at all, crop breeders drawing on wild genetic resources from widely distant geographic sources. Whether this is so or not, and whether in fact most wild genetic resources come from developing countries, can be tested by examining the origins of the wild germplasm used to improve our sample of 24 major world crops. Figure 3 shows the percentage of each crop produced in developed and developing countries; the wild species from which germplasm was obtained to improve the crop; and the country from where the wild germplasm came.

* Ten major world crops grown largely (70+%) or entirely in *developed countries* have been improved with wild germplasm: barley, oats, potato, sunflower, carrot, grapes, apple, pear, strawberry, sugar beet. The wild germplasm has come entirely from developed countries in the case of six crops (sunflower, carrot, grapes, apple, pear, sugar beet). It has come entirely from developing countries in the case of one crop (potato); and from a mixture of developed and developing countries in the case of three crops (barley, oats, strawberry).
* Eight major world crops grown largely (70+%) or entirely in *developing countries* have been improved with wild germplasm: rice, cassava, sweet potato, oil palm, sesame, sugarcane, rubber, cacao. In all eight cases, the wild germplasm has come entirely from developing countries.
* Six major world crops grown *in both* developed and developing countries have been improved with wild germplasm: wheat, maize, tomato, peas, (in mostly developed countries); tobacco, cotton (in

mostly developing countries). The wild germplasm has come from both developed and developing countries in the case of five (wheat, maize, peas, tobacco, cotton). It has come entirely from developing countries in one case (tomato).

This analysis suggests that the flow of wild germplasm that has actually been used has largely been North-North (among developed countries) and South-South (among developing countries). But there is also a significant North-South/South-North exchange in which the North is a net gainer. The disparity is not as great as some analysts have suggested but nonetheless it does exist. But this is a picture of the *status quo*. Will it change with the anticipated changes in the use of wild germplasm?

To answer this question we have looked at the distribution of all the wild relatives of the top crops in each category: wheat, rice and maize among the cereals; potato, cassava and sweet potato among the root crops; soybean, sunflower and oil palm among the oil crops; tomato among the vegetables and beans among the pulses; grapes, banana and citrus among the fruits, sugarcane and sugar beet; tobacco, coffee and rubber among the commodity crops; and cotton among the fibre crops.

Five of these — soybean (67% of the crop grown in developed countries; 33% in developing countries), beans (22% developed; 78% developing), banana (2% developed; 98% developing), citrus (51% developed; 49% developing) and coffee (100% developing) — have not yet been improved with wild germplasm but can be expected to be in the future. The remainder have been improved, and can be found in the list above. An even greater range of wild species is likely to contribute to most if not all of these crops in the future.

The wild gene pools (GP1 + GP2 + GP3) of these 20 crops can be classified as follows:
* Entirely in developed countries: none
* Entirely in developing countries: three (oil palm, tomato, rubber)
* Mostly (70+% of the species) in developed countries: two (sunflower, grapes)
* Mostly (70+% of the species) in developing countries: 12 (rice, maize, potato, cassava, sweet potato, banana, citrus, sugarcane, tobacco, coffee, beans, cotton)
* More evenly split between developing and developed countries: three (wheat, soybean, sugar beet).

From this we conclude that there will continue to be considerable exchange of wild germplasm among developed countries on the one hand and among developing countries on the other; and that the flow of wild germplasm from developing to developed countries will probably increase. Of the 15 crops whose wild gene pools are mostly or entirely in developing countries, nine are largely grown in the developing countries themselves, one (potato) is largely grown in developed countries, and five are produced by both groups. So the greater concentration of wild germplasm in the developing countries will not necessarily be reflected proportionately in the South-North flow.

This analysis is rough and ready and there are bound to be exceptions to the patterns identified here. The patterns also obscure exchanges of germplasm of potentially major effect. Developed countries produce only 7% of the world's rice, but the 2% accounted for by the US crop alone has a farm-gate value of more than a billion dollars and much of it is consumed in developing countries. Genes from the wild species have not been used to improve the US cultivars, but cultivars are being developed with disease resistance from wild *Oryza rufipogon* (found entirely in developing countries). Nevertheless we think the pattern is sufficiently clear to draw three conclusions.

First, there can be no such thing as self-sufficiency in wild genetic resources. Every country, regardless of its level of development, depends on other countries for wild genetic resources.

Second, developed countries as a group and developing countries as a group are each rich in wild genetic resources. It is often stated that developed countries are poor in wild genetic resources. The argument seems to be as follows: "Tropical environments are richest in species: tropical environments are mostly in developing countries: therefore most species are in developing countries: therefore developed countries have few wild genetic resources". It is true that exclusively temperate and subarctic developed countries (Canada, northern Europe) have few wild relatives of crops, but this is not the case with those developed countries with Mediterranean and subtropical environments.

Third, the South-North flow of wild genetic resources is nevertheless significantly greater than the North-South flow.

So far in this chapter, we have only discussed the use of wild germplasm in agriculture. This is because comparable data from which reasonably objective conclusions can be drawn are available only for agricultural crops. We doubt, however, if inclusion of data on forage crops, timber trees, or aquaculture would lead to different conclusions.

The major exchanges of wild timber and aquacultural genetic resources tend to be among developed temperate countries as one discrete group and among developing tropical countries as another. The reason for this is straightforward. Temperate germplasm is adapted to temperate conditions, and tropical germplasm to tropical conditions, and there is no advantage in attempting tropical/temperate exchanges. Pines from Central America are grown in East Africa; pines from California are grown in New Zealand; and so on. The major exception to this is germplasm from subtropical and Mediterranean-climate areas, which is suitable to a wider range of countries.

From the point of view of the production of living resources, the world can be divided into three zones: boreal (subarctic)-temperate; Mediterranean-subtropical; and tropical.

* There are only two developing countries in the boreal-temperate zone: Argentina and Chile. So most of the exchange of wild germplasm in this zone is among developed countries (exceptions being the South-North movement of wild potato and strawberry germplasm).

Figure 3. Origins of the wild germplasm used to improve 24 major world crops

CROP	% OF PRODUCTION GROWN IN		SOURCE OF WILD GERMPLASM	
	DEVELOPED COUNTRIES	DEVELOPING COUNTRIES	SPECIES	COUNTRY
Wheat	66	34	*Triticum turgidum dicoccoides*	Turkey or Israel*
			Aegilops umbellulata	Turkey*
			Ae. ventricosa	Italy or Spain*
			Agropyron elongatum	?
Rice	7	93	*Oryza nivara*	India
Maize	65	35	*Tripsacum dactyloides*	(USA-Venezuela)
Barley	87	13	*Hordeum spontaneum*	Turkey*
Oats	94	6	*Avena sterilis*	Israel, Portugal, Algeria & Tunisia
Potato	83	17	*Solanum acaule*	(Argentina, Bolivia, Peru)
			S. demissum	Mexico
			S. spegazzinii	Argentina
			S. stoloniferum	Mexico
			S. vernei	Argentina
Cassava	0	100	*Manihot glaziovii*	Brazil
Sweet potato	2	98	*Ipomoea trifida*	Mexico
Sunflower	79	21	*Helianthus annuus*	USA
			H. petiolaris	USA
Oil palm	0	100	*Elaeis guineensis*	Ivory Coast, Nigeria & Zaire
Sesame	0	100	*Sesame orientale malabaricum*	India
Tomato	60	40	*Lycopersicon esculentum cerasiforme*	Ecuador & Peru
			L. cheesmanii	Ecuador
			L. pimpinellifolium	Ecuador & Peru
			L. chmieleswkii	Peru
			L. hirsutum	Ecuador & Peru
			L. peruvianum	Peru & Chile
Peas	61	39	*Pisum fulvum*	(Israel, Jordan Lebanon, Syria, Turkey)
Carrot	74	26	*Daucus carota*	USA (introduced)

CROP	% OF PRODUCTION GROWN IN		SOURCE OF WILD GERMPLASM	
	DEVELOPED COUNTRIES	DEVELOPING COUNTRIES	SPECIES	COUNTRY
Grapes	80	20	*Vitis amurensis*	USSR
			V. aestivalis	USA
			V. berlandieri	USA
			V. lincecumii	USA
			V. riparia	USA
			V. rupestris	USA
			V. labrusca	USA
Apple	77	23	*Malus baccata*	USSR
			M. floribunda	Japan
Pear	78	22	*Pyrus sp.*	USSR
Strawberry	90	10	*Fragaria chiloensis*	USA, Canada & Chile
			F. ovalis	USA
			F. virginiana	USA
Sugarcane	9	91	*Saccharum robustum*	Papua New Guinea
			S. spontaneum	India & Indonesia
Sugar beet	92	8	*Beta maritima*	Italy
Tobacco	39	61	*Nicotiana debneyi*	Australia
			N. glutinosa	Peru
			N. longiflora	(Argentina, Bolivia, Brazil, Paraguay, Uruguay)
			N. plumbaginifolia	(As *N. longiflora* + Peru)
			N. suaveolens	Australia
Rubber	0	100	*Hevea brasiliensis*	Brazil
Cacao	0	100	*Theobroma cacao*	Peru
Cotton	42	58	*Gossypium anomalum*	(Cameroon, Central African Rep, Chad, Ethiopia, Niger, Nigeria, Somalia, Sudan, Angola & Namibia)
			G. hirsutum mexicanum	Mexico
			G. tomentosum	USA

Source countries which are asterisked are the most probable source, but are not known for certain; those in brackets are where the source country is not known, but the species occurs wild in the country concerned. *Developed countries* include all countries in Western and Eastern Europe, plus the USSR, Canada, the USA, Australia, New Zealand, Israel, Japan and South Africa; *developing countries* include all other countries.

* There are only two developed countries in the tropical zone: Australia and the US (Hawaii + southern Florida). So most of the exchange of wild germplasm in this zone is among developing countries (exceptions being the South-North movement of wild potato and tomato germplasm).
* The Mediterranean-subtropical zone is entirely different. The US, the USSR, Japan, Australia, New Zealand, South Africa, Israel, and the nations of southern Europe all have substantial areas of plant and animal production in this zone. So have many developing countries, such as China, India, the countries of West Asia and North Africa, Brazil and Argentina.

Both developed and developing countries in the Mediterranean-subtropical zone are rich in the wild relatives of crops. The USSR, particularly Transcaucasia (Georgia, Azerbaijan and Armenia), has many wild relatives of wheat, rye, barley, apple, pear, plum, apricot, almond, cherry, currant, grape, walnut, pistachio and alfalfa (190). The US contains the centres of diversity of sunflower and grapes, as well as wild strawberries, hops and cotton and the northernmost relatives of potato, cassava, sweet potato, maize, beans and squashes (137). Australia is home to wild tobacco, soybeans, cotton, rice, citrus and banana (118).

It is no accident that the greatest amount of exchange of wild germplasm between developed and developing countries relates to gene pools of the Mediterranean-subtropical zone and of the interface between this and the other two zones. The Mediterranean basin itself is a classic example, the north developed, the south developing, but both sides enjoying an abundance of wild species related to wheat, barley, oats, chickpeas, lentils, peas, grapes, figs, olives and forage plants such as clovers and medics.

Even when countries *appear* to be giving more wild genetic resources than they receive this may not always be the case. The global exchange of germplasm is complicated by the fact that germplasm is exchanged in various forms at various stages of development: from "raw" wild germplasm through breeding lines to advanced cultivars. The breeding lines and cultivars may contain the germplasm of a wild species from one country, genes from traditional cultivars (landraces) from several other countries, and genes from other advanced cultivars from yet other countries.

The movement of wild germplasm from developing to developed countries is not necessarily a one way street. The germplasm can return to the developing country reconstituted as a new or better cultivar. Australian forage crops illustrate this type of exchange. *Neonotonia wightii* is a perennial forage legume, native to eastern tropical and subtropical Africa, the Arabian peninsula and India. In Australia, collections of wild plants from Kenya, Tanzania, South Africa and Malawi were developed into the cultivars Tinaroo, Cooper, Clarence and Malawi respectively (9; 115). These have proved so successful that they have been widely adopted in Africa and Latin

America, particularly Brazil. Live weight gains of cattle in Brazil on a mixture of *N. wightii* and pangola grass (*Digitaria decumbens*) were 316 grams (3.7oz)/day/head, compared with 231 grams (2.7oz)/day/head on unfertilized grass alone and 294 grams (3.5oz)/day/head on fertilized grass alone (17).

Macroptilium atropurpureus, another perennial forage legume, native from the US (southern Texas) to Argentina, has a similar history. A collection of wild plants from Mexico was developed into the Australian cultivar Siratro. The cultivar has been introduced to Kenya, Zimbabwe, India, Jamaica, Cuba, Brazil, the Philippines, Papua New Guinea, Fiji and the US. It is highly valued for its reasonable drought tolerance, capacity to withstand mild frosts, good nitrogen fixation, and productivity. It is now one of the most popular leguminous pasture plants in the tropics and subtropics (17).

The exchange of wild germplasm, like the exchange of domesticated germplasm, is rather informal, governed only by tacit recognition of two principles first promulgated by FAO:
* Material will be freely and fully available to all who can make use of it for the benefit of humanity.
* Duplicates of material collected are always left in the country of origin.

Opinions differ about how well this system is working, and in recent years there have been calls to formalise the rights and responsibilities of nations concerning the conservation and exchange of genetic resources. These calls were crystallised into a request by the FAO conference in 1981 that the director-general of FAO draft an international convention for the free exchange of plant genetic resources, and study the possible establishment of an international FAO gene bank.

A background document on the proposed international convention and gene bank was prepared by the FAO secretariat, and discussed by FAO's Committee on Agriculture in April 1983. Several of the member countries on the committee doubted the need for the convention, or doubted that any convention could guarantee free exchange. The committee decided that the document should be redrafted and be submitted to the FAO conference in November 1983.

An international convention may still emerge in due course, but the international gene bank appears to be a nonstarter. Its establishment and management would be extremely expensive, which makes it highly unattractive to governments in the present recession. In any case, it would have to be restricted to seed-propagated crops; vegetatively propagated crops have to be maintained in areas to which they are adapted ecologically (133a).

Three factors appear to be behind the call for an international convention and gene bank. First, there is a belief that developing countries are being ripped off: that they have all the genetic resources, but the developed countries are the only ones with the money and research and development (R & D) skills to make use of them. Second, there are evident cracks in the current system.

Some countries have started imposing restrictions on the movement of germplasm; others are threatening to do so. Third, there is a trend towards privatisation of genetic resources: cultivars are already being patented. The use of genetic resources is now big business, and companies want to own the resources they use.

The feeling that developing countries are getting a raw deal is expressed only partly in the view that most wild genetic resources are in the South, most uses of them in the North, a view which is not supported by the facts, as we have shown. There is also the feeling that no matter where wild germplasm originates, it ends up in the hands of the North because developed countries have the gene banks and the R & D facilities.

This was certainly true a decade ago, but things have changed rapidly since then. Most of the medium and long term storage facilities for rice, maize and food legumes are now in developing countries (see Chapter 6). All the major cassava collections, and most of the major sweet potato collections, are in developing countries. Most of the major potato collections are in developed countries, but then so is most potato production (40).

The majority of gene banks in developing countries, however, are still only capable of medium and short term storage. Most of the long term storage facilities are in developed countries. Thus the developed countries have all five long term collections of wheat, both long term collections of millet, both long term collections of soybean, three of the seven long term collections of rice, and two of the three long term collections of maize, two of the three for sorghum, seven of the eight for barley and one of the two for potato (134). This is changing quickly, however, and by 1985 developing countries are expected to overtake the developed countries in the number of gene banks with long term storage capacity (134).

Unfortunately, there are unmistakable signs that the principle of free exchange of germplasm is at risk. Ethiopia has imposed a ban on the export of wild and domesticated coffee germplasm. Brazil has reported refusals by the countries concerned to supply wild oil palm from West Africa, wild black pepper from India, and castor beans. The US has made threatening noises about withholding germplasm from countries of which it currently disapproves. Some refusals to supply germplasm seem to be attributable to the feeling that germplasm exchange means giving something for nothing. Other refusals appear to be overtly political. Would an international convention, plus or minus a gene bank, solve the problem?

Participants in the FAO/UNEP/IBPGR International Conference on Crop Genetic Resources (scientists and other experts from 31 developing and 19 developed countries, and from several international agricultural research centres and other international organisations) were asked this question when they met in April 1981. The consensus was that it would not. The current informal arrangements between IBPGR and individual countries are not ideal but the alternatives seemed worse.

At the conference Mr RH Demuth, a lawyer and first chairman of IBPGR, said that the operations and characteristics of the many national and international germplasm centres involved are so diverse that it would be very difficult to accomodate them in a single convention. While a convention is being negotiated, a lengthy and time-consuming business at best, centres would be unlikely to enter new agreements with IBPGR, and the current momentum would be lost. The very existence of a legally binding international convention might prevent the regeneration of that momentum by politicising a system that has been so successful because it has been voluntary and nonpolitical, argued Demuth (50).

In fact the principle of free exchange does have a certain moral force, and the international organisations involved are not totally bereft of means to reinforce it. IBPGR's response to Ethiopia's refusal to release coffee germplasm is to say that if the country does not live up to international principles it cannot expect international support (88). Restrictions imposed by Sudan on the export of *Acacia senegal* (the tree that produces gum arabic) germplasm were lifted at the insistence of FAO. Indeed forestry officials with whom we have discussed the convention are emphatic that trees should not be included. One of them told us that trees are not "important" enough to have become a political football. Tree germplasm exchange is free, the only problem being that it is often simply unavailable, there having been too few collections of material. Similarly, Dr JT Williams, executive secretary of IBPGR, reported that exchange of the genetic resources of *food* crops is not a problem; the difficulties being encountered largely concern the germplasm of industrial and *commodity* crops (50).

Which brings us to the issue of privatisation. Private ownership poses several potential threats to the free exchange and responsible conservation of genetic resources. First, the patenting of cultivars could lead (opinions are divided as to whether it will) to restrictions on the exchange and use of the genetic resources within those cultivars. Some say that patenting restricts only the right to grow and sell the crop produced by that cultivar. Others say it also restricts the right to use the germplasm of that cultivar. Biology favours the former opinion: you can patent a particular combination of genes, provided you can show that the combination is a human artifact and that you are the human responsible, but you cannot patent the genes themselves.

A second threat is a byproduct of plant breeders' rights: the insistence by some governments, notably those of the European Economic Community, on standardisation and uniformity of cultivars. This poses a direct threat to the conservation of traditional cultivars (for example, of vegetables) in the countries concerned, since they are generally not all standard or uniform. It does not threaten the conservation of wild genetic resources, since they are not cultivars, but it could restrict some uses of them. For example, one way of dealing with the problem of proliferation of new types of disease, immune to the resistance genes in conventional cultivars, is to develop multiline cultivars:

cultivars consisting of different lines with different resistance genes which can be juggled in various mixtures depending on the combination of disease-types affecting the crop. By definition, such multilines are not uniform, and are therefore unpleasing to the bureaucratic mind.

A third potential threat posed by privatisation is that it can restrict and sometimes throttle completely the exchange of both breeding material and scientific information. Evidence for this comes from those crops in developed countries that are dominated by commercial businesses, such as sunflower, maize, cotton and many ornamental crops (flowers). In the case of sunflower and cotton, exchange of breeding material and scientific information between public and private breeders is good. With many ornamental crops (zinnias, for example) the private breeders shroud their operations in secrecy. We suspect that for as long as there are significant public breeding programmes for a crop then there will be reasonably open exchange of materials and information, since private industry stands to gain from such arrangements. But in crops where there is no significant public contribution, and so no breakthroughs by the public sector in transferring wild germplasm into usable breeding lines, then the companies' incentive to exchange data and germplasm is much reduced.

Finally, it is possible that certain commercial uses of wild genetic resources could lead to polarisation of the North-South issue. We have noted that the wild genetic resources of crops grown in the South also occur in the South. A developing country with wild genetic resources of rubber, sugar or some other tropical commodity may not grow that commodity but at least the resources are being kept within the developing country family as it were. The emerging capacity of industries to extract chemicals directly from plant cells could conceivably lead to the wholesale migration of actual or potential tropical crops from developing country plantations to developed country laboratories. For example, the sweetener thaumatin, isolated from the West African plant *Thaumatococcus danielli*, is unlikely to lead to the development of a new plantation crop since it is being "grown" successfully by bacteria. The "sweetness" gene has, by genetic engineering, been removed from the plant and placed into a bacterium.

5. Threats to wild genetic resources

Wild genetic resources are vulnerable to the same kinds of threat as are wild species. The main problems are loss of habitat, overexploitation, and competition and predation by introduced species.

But with wild genetic resources, the impacts of these threats are more difficult to detect. Valuable gene pools of widespread species may disappear undetected, either because not enough is known about the distribution of genetic variation within the species, or because the very abundance of the species masks the disappearance of its constituent gene pools.

Because of the difficulty of detection, what we know of the threats to wild gene pools is likely to be just the tip of the iceberg.

Cereals

There is good news and bad news. The good news is that some weedy relatives of rice, maize, barley, sorghum and oats are probably doing better than before. Since they flourish in disturbed ground, a habitat that is expanding rapidly as people fell forests and create zones of wasteland along roads and among buildings, the range of some of them is also enlarging. The *arundinaceum* race of wild sorghum, for example, is spreading as the tropical forests of Africa are cut back. And some of the *Tripsacum* species, GP2 and GP3 relatives of maize whose natural habitat is bare rocky cliffs, are probably increasing as the new superhighways of Mexico and Central America are blasted through the mountains (69).

The bad news is that not even the capacity to take advantage of disturbance is a guarantee of survival. The range of the closest wild relatives of maize, *Zea mays mexicana* and *Z. m. parviglumis* (both GP1), has been halved since 1900 (181). The two subspecies occur as weeds in maize fields and along fencerows, and are truly wild in only a few areas. Whether weeds or wild they exist only as scattered populations in Mexico and western Guatemala, and even the largest populations are under pressure from intensification or agriculture (cleaner cultivation) and overgrazing by livestock (181; 182). Since 1960, the rate of extinction has been accelerating (181).

We regard the wild subspecies of maize as vulnerable (139), but three other wild relatives should probably be considered endangered. *Zea perennis* (GP1), until recently thought extinct, and the recently discovered *Z. diploperennis* (GP1), both have highly restricted distributions in the state of Jalisco, Mexico

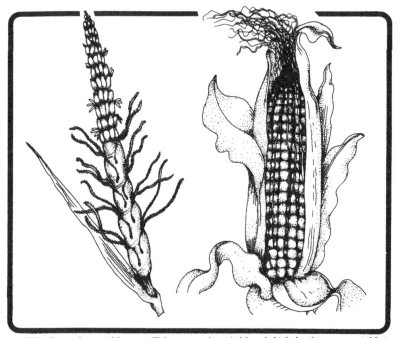

*MAIZE. Genes from wild grass (*Tripsacum dactyloides: *left) helped increase yields in widely-grown American and Mexican maize hybrids in the late 1960s and early 1970s. The greatest future promise however lies with wild maize species more closely related to cultivated maize (right).*

(42; 93). And *Z. luxurians* (GP1) of southeastern Guatemala, which of all the annual relatives of maize is most threatened with extinction in its native habitat, survives only where it can escape grazing pressure along fencerows and among rock outcrops (182).

Overgrazing also threatens wild gene pools of wheat, oats and sorghum. Through much of its range in West Asia, overgrazing has reduced the wild wheat *Triticum turgidum dicoccoides* (GP1 of macaroni wheat; GP2 of bread wheat) to rocky or stony habitats where it can escape grazing (192). Where grazing is controlled, "non-arable sites support stands as dense as cultivated wheat fields", said a 1966 report (70). One form of wild sorghum has virtually disappeared from some sites due to extreme overgrazing (67).

According to the oat expert Dr Bernard Baum, overgrazing is seriously affecting some important areas of wild oat genetic diversity, notably among the species *Avena clauda*, *A. eriantha*, and *A. ventricosa*. However, in North Africa and West Asia where the first two of these species are native, economic development has provided alternatives to pastoralism as a source of income and employment. So grazing pressure is diminishing due to abandonment of livestock raising for other economic pursuits, and these two species are showing signs of recovery (11). This is an example of development promoting conservation.

A number of wild wheat and oat species are confined to rather small areas, and so are intrinsically vulnerable. *Aegilops comosa* (GP3 of wheat) occurs only in Greece and west Turkey; *Aegilops vavilovi* (also GP3) only in Syria; and *Triticum timopheevi araraticum* (GP2) in scattered parts of Iran, Iraq, Turkey and the USSR. *Avena canariensis* (GP2 of oats) is found only in the Canary Islands and a small area of Morocco facing it; *A. damascena* only in Syria; *A. macrostachya* only in Algeria; and *A. maroccana* only in Morocco (all three GP2). Urban and agricultural expansion, and other changes in land use, are particularly threatening to such restricted species.

Gene pools of wild rice (*Oryza*) species are also at risk, and some have disappeared altogether. Deforestation has depleted wild *Oryza* habitat in southern China (180), and a form of wild *O. rufipogon* in the Taoyuen district of Taiwan disappeared from its restricted habitat during the 1960s (28). Irrigation, drainage, highways, industries and housing, as well as the intensification of agriculture which discourages the persistance of weedy relatives such as *O. rufipogon* and *O. nivara* (both GP1), have destroyed populations of these two species and threaten those of others (28; 29).

Root crops

More than half the wild species in the genera *Solanum* (potato) and *Manihot* (cassava) are narrowly endemic (have highly restricted distributions) in South, Central and Middle America. The two main centres of diversity of the 154 wild species of potato are in central Mexico and the central Andes of South America. Among the 98 wild cassava species, the main centres of diversity are east-central Brazil (southern Goias and western Minas Gerais) with 38 species, northeastern Brazil with 18 species, and southwestern Mexico with 16 species (127; 148).

David J Rogers and SG Appan, authors of the monograph on *Manihot* (148), believe that most of the cassava species in east-central Brazil are of recent origin, having developed after fires started by people had burned the vegetation at frequent intervals. Even if this is so, there are signs that this diversity may now be decreasing. Nagib MA Nassar, Professor of Genetics of the Institute of Biological Sciences at the Federal University of Goias in Brazil, reports the disappearance of *Manihot glaziovii* and other wild *Manihot* species from several areas, due to conversion of land to agriculture and to extermination of the plants by livestock raisers. Most of the *Manihot* species are poisonous to grazing animals, due to high levels of cyanide-producing chemicals (cyanogenic glycosides), so farmers do their best to get rid of them (127).

Oil crops

Soybean (*Glycine max*) has one wild relative, (*G. soja*) in its primary gene pool, no secondary gene pool, and no tertiary gene pool, since all attempts to cross the six other wild species of *Glycine* with *G. max* have been unsuccessful (85). The habitats of *G. soja* in Japan, Korea, Taiwan, northern and central China, and adjacent areas of the USSR, are being destroyed by roads, houses, factories and airports (83; 84; 85). *G. latrobeana*, one of the species that may eventually be included in the tertiary gene pool, is classified as vulnerable by the Australian National Parks and Wildlife Service (107). The species occurs as scattered populations in open grassland pastures in only three states in Australia (South Australia, Victoria, and Tasmania); some of the populations have already been lost to urban sprawl and grazing (85; 107).

Although some wild sunflowers, notably wild forms of *Helianthus annuus*, are widespread and flourish under disturbed conditions, others are confined to a few localities and are near extinction. The serpentine sunflower, *H. exilis* (GP1), is an endangered species restricted to a small part of California (US), which suffers from habitat disturbance and competition from more tolerant sunflower species. Yet the species has considerable agronomic potential, because it has the highest linoleic acid (a key fatty acid) percentage of any sunflower species examined so far, and is a source of cold tolerance and genetic fertility restoration (95; 149).

Another Californian sunflower, *H. nuttallii parishii* (GP2), may already be extinct. Its only known habitat has been destroyed by the expansion of Los Angeles, and despite several efforts it has not been found for some years (149). The paradoxical sunflower, *H. paradoxus* (GP1), is yet another endangered species. Only four populations are known, one of which was completely destroyed during construction of a highway, and another by the widening of a bridge (149). Eight other sunflower species are considered rare. But important characteristics such as disease resistance are not distributed uniformly throughout the species, and gene pools of particular value may be at risk (137). So even the abundant species are in need of conservation.

The wild oil palm (*Elaeis guineensis*) is a notable example of a species that has prospered from habitat disturbance. It is a plant of the forest edge, intolerant of shade, and as the West African forests have been pushed back by settlement and agriculture the oil palm has followed. The species became the dominant component of secondary growth. Its survival was assured even when the vegetation was repeatedly cleared and burned in the course of shifting (slash-and-burn) cultivation, because of its usefulness as a source of oil and wine.

Two new developments, however, now threaten some of the wild gene pools of the oil palm. One is clearance of large tracts of land for urban expansion, industry, and modern plantation and monoculture-type agriculture. The other, a direct consequence of population expansion, is the thinning of palm groves. Human pressure on the land has led to increasingly extensive thinnings, so that more food can be grown. This can reach the point where so few palms remain that they are not worth keeping, and before people realise what has happened another grove and another gene pool have disappeared (5).

Experts differ over the severity of these threats to the African oil palm (*E. guineensis*). Dr Arasu and Dr Rajanaidu of the Malaysian Agricultural Research and Development Institute regard several Nigerian groves as close to the point of no return. They say that if a similar situation exists in the other oil palm countries of West and Central Africa, "then the collection and conservation of oil palm germplasm will change from something that might usefully be done to something that urgently needs to be done" (5). But Ir G Blaak, senior agronomist at the Royal Tropical Institute in the Netherlands, states: "There is no danger that the wild African oil palm will be extinct" (16a).

By contrast, there is general agreement on the status of the wild American oil palm (*E. oleifera*). *E. oleifera* has great potential in the improvement of the African oil palm, because it can provide short stature (shorter trees are easier to harvest), higher content of unsaturated fatty acids (*E. guineensis* oil is 50% unsaturated; *E. oleifera* oil is 80% unsaturated), adaptations both to very wet and very dry soils, disease resistance (for example to vascular wilt caused by the fungus *Fusarium oxysporium*, common in West Africa), and pest resistance (for example, to leaf miner *Coelaemenodera elaeidis*, very destructive in West Africa) (0a; 141a). Unlike its African relative, the American oil palm has little direct economic value, and so is not spared when land is cleared. As it occurs in small groups rather than large groves, its component gene pools are easily eliminated. Although its range is quite large (Central America and northern South America), its populations are scattered and vulnerable to habitat destruction. Because of its importance and vulnerability, *Elaeis oleifera* is one of the three species that we have recommended as a priority for *in situ* conservation by the International Board for Plant Genetic Resources (IBPGR) (138).

The other two species are wild *Coffea arabica* (discussed under commodity crops), and *Olea laperrinei*, a close relative of the olive (*O. europea*). This wild olive grows in the isolated mountains of Sudan, Niger and southern Algeria. The International Union for Conservation of Nature and Natural Resources (IUCN) regards it as vulnerable, since the population in Niger and Algeria are reportedly not regenerating, due to a combination of dessication and the pressure of grazing, browsing and cutting. The young branches of the trees are apparently a favoured source of cattle fodder (112). The populations in Sudan are in much better condition. A UNDP/FAO report has recommended that a nature reserve be established on the western slopes of the Jebel Marra, which would protect good stands of this wild olive (112).

Vegetables and pulses

All the tomato's wild relatives have limited distributions, which makes the crop's wild gene pools exceptionally prone to erosion by habitat destruction. One of the GP1 species, *Lycopersicon cheesmanii*, is protected in the Galapagos Islands National Park; but the habitats of others are being obliterated by agricultural, industrial and residential expansion. Dr Charles Rick, the world's foremost expert on wild tomatoes, reports that certain stands of *L. hirsutum* and *L. peruvianum* known 20-25 years ago no longer exist (144).

Many chickpea (*Cicer*) species are also threatened. Their populations are scattered, sometimes far apart, often consisting of only a few individuals. The perennial species are slow to establish themselves and do not produce much seed. They are particularly vulnerable to habitat destruction and overgrazing. The species most at risk are probably *C. echinospermum* (GP2), confined to southeast Turkey; *C. floribundum*, found only in southern Turkey and threatened by removal of its forest habitat; and *C. graecum*, perhaps the most threatened of all, since it is limited to one small population in Greece.

Others may be secure at present as species, but some of their populations are at risk: *C. anatolicum* (Turkey, Iran, Iraq, USSR); *C. bijugum* (Turkey, Syria, Iraq); *C. microphyllum* (Afghanistan, Pakistan, India, China); *C. montbretii* (Turkey, Bulgaria, Albania); *C. oxydon* (Iraq, Iran, Afghanistan); *C. reticulatum* (GP1) (Turkey); *C. spiroceras* (Iran); *C. tragacanthoides* (Iran, USSR). Still others have such a highly restricted distribution that they may well be threatened, but not enough is known of their status to say whether they are or not: examples are *C. atlanticum* (in the Atlas Mountains, south of Marrakesh, Morocco); *C. baldzhuanicum* (South Tadzhikistan, USSR); *C. stapfianum* (only one site known, near Shiraz, Iran); *C. yamashitae* (near Kabul, Afghanistan) (117b; 116).

Fruits and nuts

Shifting cultivation and logging are threatening the variability of wild bananas (*Musa acuminata* (GP1) and other *Musa* species: 86). In Kalimantan, Indonesia, *Musa* species that are part of the second storey vegetation of the tropical rainforest are being destroyed by logging operations. They are often replaced by other kinds of wild bananas that flourish in the full sunlight of logged-over areas, but these are the very areas that are preferred by shifting cultivators for the establishment of new fields, so these species disappear too (175).

The wild relatives of citrus fruits are also losing genetic diversity due to the reduction of Southeast Asia's tropical rainforests (158). The citrus group (*Citrus, Clymenia, Eremocitrus, Fortunella, Microcitrus* and *Poncirus*) has a high proportion of species with restricted distributions, some of which have been discovered only recently. *Citrus halimii*, wild in primary tropical rainforest peninsular Malaysia and Thailand, although discovered in 1902 was not identified as a new species until some 70 years later (167).

Another new species, *Microcitrus papuana*, was discovered in 1970 in the Central District of Papua New Guinea, in the transition zone between rainforest and coastal savanna. So far, its only known location is an area that has been heavily logged and where cleared land is being planted with teak (*Tectona grandis*). The new species is being tested as a possible rootstock for citrus fruits (185). These two examples illustrate the possibility of finding new species of potential genetic value, but the possibility diminishes as the areas of greatest ecological diversity shrink before the advance of logging, settlement and agriculture.

The wild gene pools of many other fruit and nut crops are being reduced throughout the world, by habitat destruction, cutting for fuel, and overgrazing and overbrowsing by domestic animals. An example from a developed country are the wild pears of Japan. The wild Japanese forms of *Pyrus ussuriensis* and *P. pyrifolia*, and the wild species *P. dimorphophylla*, are of great importance as genetic resources for the improvement of Chinese and Japanese pears. But, according to Dr Muneo Iizuka of the Faculty of Horticulture, Chiba University (Japan), they are "headed towards extinction" (92). Only *P. dimorphophylla* (Mamenashi) has been given any protection. The others are losing their habitats to urbanisation and industrial development. Aonashi, a wild form of *P. ussuriensis* (sometimes given specific status as *P. hondoensis*) growing as scattered individuals and in small populations in forested areas, is rapidly disappearing and is now only rarely found (92).

Sugar crops

Valuable germplasm of wild sugarcane (*Saccharum*) and related species has apparently already been lost as a result of urbanisation and agricultural expansion in Malaysia, Indonesia and Papua New Guinea, the area of greatest diversity (41). Such pressures, combined with forestry and mining developments, continue to erode the region's wild sugarcane gene pools (15). Although the species are generally resilient and many of them are widespread (notably *S. spontaneum*), the various geographic groups are distinctly different genetically and representative populations need conservation.

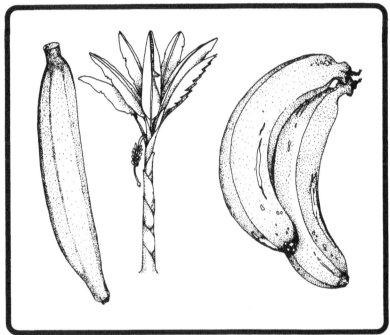

*BANANA. Disease resistance from wild bananas (*Musa acuminata: *left) has been transferred by breeders to cultivated bananas (right). But another ten years work is probably needed before the quality of the new strain is up to commercial levels.*

Commodity crops

Wild and semi-wild coffee, *Coffea arabica* (GP1), in Ethiopia is endangered by removal of the last remaining forests in the southwestern part of the country. The main causes of the forest destruction are agricultural expansion and resettlement of people from the eroded, impoverished highlands. Seven-eighths of the forest cover of Ethiopia had vanished by the mid-1960s and the remainder is under heavy pressure (53; 55). An additional threat to many of the semi-wild and primitive domesticated forms is the introduction of coffee berry disease (CBD), caused by *Colleotrichum coffeanum*. Most genotypes are susceptible to the disease, and farmers are likely to replace them with newly developed resistant cultivars (88). Wild *Coffea arabica* has been designated by the IBPGR as a priority for *in situ* conservation.

Cocoa and chocolate are both made from the beans of the cacao tree, *Theobroma cacao*. Wild and semi-wild germplasm (GP1), has already disappeared and more losses are feared. The centre of diversity of the species is northeastern Peru and adjacent Ecuador, a region of resettlement, logging and oil exploitation. Very rapid changes have been taking place in the Oriente region of Ecuador, as a result of the discovery of important oilfields. Large areas have been bulldozed to make way for well sites, field camps, airstrips and other facilities. As the region is opened up and roads are driven through the forest, new settlements are built and agricultural colonisation schemes set up (27). This area has already provided cacao growers with wild and semi-wild germplasm that has revolutionised cacao breeding and resulted in cultivars with better disease resistance and higher yields. However, the genetic base of the crop is still narrow, new sources of disease resistance are badly needed, and the richest reservoir of untapped genetic material is the centre of diversity, which even today has been largely unexplored and little collected by cacao scientists (27; 89; 162).

Fibre crops

Of the 34 wild species of cotton (*Gossypium*), 22 have highly restricted distributions, and the wild forms of the two main cultivated species (*G. barbadense* and *G. hirsutum*) are also confined to small areas. *G. klotschianum* is rare but protected in the Galapagos Islands National Park (136). We regard *G. raimondii* as endangered: it has been found only in the departments of Cajamarca and La Libertad in northern Peru. It is reported by Harland (72) to be "extinct in the area in which it was first found", and by Leon (109) to be "reduced now to a few individual plants in one or two valleys".

Timber

A number of species that once had value as sources of timber are now close to extinction. An example is the tarout cypress (*Cupressus dupreziana*), confined to a 200 square kilometre (124 sq miles) area of the Tassili N'Ajjer massif in eastern Algeria. The species is one of the most drought-resistant known, and is also highly tolerant of frosts. It produces a fine, aromatic timber of medium density that has been described as suitable for the most exacting uses. At last count there were only 153 living individuals but many more dead ones. The species was heavily used as a source of firewood by nomadic tribespeople in the area, whose livestock also destroyed any regeneration. Firewood is still taken from the remaining plants, but there are so few of them now and they are so scattered (less than one tree per square kilometre: 1.6 trees/sq mile) that they have ceased to be a significant source of fuel. Assuming the surviving germplasm can be saved, the tarout cypress has good potential as a plantation species and as a source of genes for the improvement of other cypresses.

Even the dead trees are valuable. *C. dupreziana*, like *C. atlantica*, another North African cypress that is also threatened, is (or was) very longlived, reaching ages of more than 2,000 years. As such the dead stumps have considerable scientific interest for dendrochronology: using tree rings to study and date past climatic events. However, the dead wood is also being collected for fuel and is in danger of disappearing completely (49; 112).

There are many more wild species of current value for their timber that are still fairly secure as species but are losing important gene pools. Examples include the *Araucaria* pines and some of the major *Eucalyptus* species. The hoop pine *Araucaria cunninghamii* occurs on the island of New Guinea and in northeastern Australia. It is an important tree for Papua New Guinea's saw log and plywood industries and also has potential as a source of woodpulp and chips. In performance tests of different populations, hoop pines from New Guinea have consistently been found to produce more and better timber than populations from Australia.

The New Guinea populations are also expected to be better adapted to conditions in other tropical countries than are the ones from Australia. Unfortunately a number of the New Guinea populations have been depleted and some have been almost completely destroyed. There are three reasons. First, there is growing demand for land on which to grow food. Second, there is fire, whether started naturally, or by people burning areas to clear land for farming or to promote grass for hunting. Fires reduce the vigour and regenerative capacity of the trees they do not kill, and the loss of vegetation greatly increases soil erosion. Third, there is excessive logging. In some areas the timber industry has virtually eliminated large stands, leaving too few survivors to assure adequate regeneration. Sometimes the logged areas are taken over for gardens, leaving still fewer trees to survive and reproduce. Once a population is reduced to a remnant by competing land uses, fire or extractive logging, it is extremely vulnerable to destruction by termites and by domesticated and feral pigs. The pigs alone can completely prevent regeneration once a stand is small enough (49).

The klinkii pine (*Araucaria hunsteinii*) occurs only in Papua New Guinea and faces problems similar to those encountered by its slightly more widespread relative, the hoop pine. It is regarded as a better plantation tree than the hoop pine, and is highly valued for its timber, which is excellent for superior plywood, flooring and interior work such as moulding, joinery and cabinets. Like the hoop pine it is not endangered as a species, but several of its scattered populations are at risk, due to shifting agriculture, fire and imprudent logging. Several stands have been reduced to a quarter of their original size, and the pigs are getting ready to finish the job (49).

Eucalyptus deglupta is one of the most widely planted eucalypts in the tropics. Native to Mindanao (Philippines), Sulawesi, Ceram and Irian Jaya (Indonesia), and Papua New Guinea, it is one of the only two species of large eucalypt that is naturally adapted to the humid tropics. Because of its rapid growth rate and excellent wood properties it is expected to become a major source of industrial fibre in the wet tropical regions of Africa, Central and South America, and Asia and the Pacific. Unfortunately its natural habitat is on the fertile and deep soils of river flats, ideal sites for agriculture. Especially in Indonesia and the Philippines, many stands are being logged for building timbers and firewood, and the sites converted to food production. One of the best provenances for the establishment of future plantations, and in fact a seed source of the Bislig Bay Lumber Company, occurs on the Caliawan River in Mindanao. More than 60% of this stand has been cleared for agriculture. The destruction is continuing, and the stand is expected to be cleared completely unless it is reserved for seed-collection (49).

Forage crops

Some species, no doubt a minority, do better as a result of human activity. Others, no doubt the majority, do worse. Although it is clear that the changes people are wreaking on the biosphere will result in a net species deficit, it is equally clear that the notion of "humans in: other species out" is an oversimplification. As we have already noted, frequent burning of vegetation may have promoted speciation in cassava (*Manihot*); and species naturally adapted to colonising disturbed ground have expanded and prospered with the spread of human disturbance. Forage crops provide some interesting examples of the ebb and flow of wild gene pools, and show how difficult it is to establish that human impact on particular associations of gene pools and ecosystems is unequivocally good or bad.

Australia has been the proving ground of some major forage crops, most notably subclover (*Trifolium subterraneum*) and the annual medics (*Medicago* species). These species were of no practical economic consequence in their native lands, and their valuable agronomic qualities were discovered and developed only after their long passage to the southern hemisphere. In the course of these developments, three trends have emerged: displacement of native Australian pasture species by the introduced Mediterranean species; differentiation of new genotypes in at least one of the introduced Mediterranean species (the first step in speciation: the development of a new species); and displacement of native pasture species in the Mediterranean by superior Australian cultivars of the same species.

As far as we can tell, the first of these trends is the one that has received least comment. The superior performance of the introduced pasture species is so evident that there seems to be no point in worrying about the erosion of the less productive Australian perennials. However, just as the performance of the Mediterranean species was no guide to their performance in Australia, so the utility of the Australian species cannot be judged from Australian experience alone. Among those species there may be adaptations of great value to livestock production in other countries. Some potentially useful species, such as the soybean relative *Glycine latrobeana*, are known to be indirectly threatened by the success of the Mediterranean species; important gene pools of other species may also be threatened.

The second trend, the emergence of new genotypes of subclover in Australia, is of great evolutionary interest and may be a source of new genetic combinations for breeding. In a study of 1,285 samples of the *subterraneum* subspecies of *Trifolium subterraneum*, Cocks and Phillips (32) discovered 435 new and undescribed strains. These new strains diverged from the original varieties in several significant respects: in appearance, such as hairiness of leaves, combinations of leaf marks, flower colour, and seed colour; in behaviour — for example, flowering earlier or later than the original strains; and chemically, such as in the content of isoflavones (oestrogens). In one pasture sown 20 years earlier with 26 named lines, there were now 332 distinct lines. The new strains are widespread, and indicate that the *subterraneum* subspecies is evolving in Australia. Probably this is a result of a combination of occasional cross-fertilization (*T. subterraneum* is normally self-fertilizing) and mutation. (No such changes have been observed in other subspecies.)

Meanwhile, back in the Mediterranean, the success of the Australian cultivars of subclover and the annual medics is catching on. If you are raising livestock in Italy, why settle for the modest productivity of natural pasture when there is a source of high productivity readily available and perfectly adapted to local growing conditions? Hence the third trend. Australian farming systems are being adopted in the lands in which subclover and the annual medics are native, and as the frequency of the introduced Australian strains increases so it is likely that the native genotypes will become more rare. For example, Jemalong, an Australian cultivar of *Medicago truncatula*, is very successful in Iraq and Algeria (33a). Such substitution threatens further progress in the development of cultivars, since it could cause the disappearance of genetic resources not yet incorporated in the crop: for example, better resistance to *Sitona* weevil in *Medicago*; or the special soil adaptations plus low levels of oestrogens of the *yanninicum* subspecies of subclover, rare both in Australia and its countries of origin.

The Mediterranean legume species are not only being replaced by exotic cultivars; they are also being heavily overgrazed by goats and sheep. In their monograph on *Medicago* (110), Dr Karlis Adolfs Lesins and his late wife, Irma, describe how in a number of Mediterranean countries they saw "only too often, depleted, overgrazed areas where some centuries ago the ground had been well covered with vegetation, often forested. On the cleared land herds of goats and sheep snap off every edible shoot so that the land becomes desert-like. The goats devour not only green shoots but search out ripe *Medicago* pods fallen to the ground." It is not possible for even quite robust populations to withstand such persistent pressure.

Livestock

As we remarked in Chapter 2, the close relatives of livestock are generally worse off than the relatives of crops. In the cattle genus *Bos*, two of the four surviving wild species are classified by IUCN as endangered and the other two as vulnerable (170). Of six sheep species (*Ovis*), one is vulnerable and one subspecies endangered. One of the four pigs (*Sus*) is endangered. Out of six goat (*Capra*) species, one is endangered, one vulnerable, and one subspecies is endangered. Three of the four Asiatic buffalo (*Bubalus*) species are endangered; the fourth is vulnerable. Of the seven surviving *Equus* species (horses and donkeys), two are endangered and three are vulnerable (170).

The birds are doing better than the mammals. None of the four junglefowl (*Gallus*) species, the chicken's closest wild relatives, is considered threatened; nor is the wild turkey (*Meleagris gallopavo*) or its cousin the ocellated turkey (*Agriocharis ocellata*). One of the 37 or so species (authorities differ on the exact number) in the *Anas* (duck) genus is vulnerable, two subspecies are also vulnerable and one is endangered; and one of the two species in the *Cairina* (muscovy duck) genus is vulnerable (102).

The most common threat to these 25 species and subspecies of mammals and birds is overhunting. Hunting is listed as a threat to every one of them, except Przewalski's horse (which may be extinct in the wild) and two of the duck subspecies (*Anas platyrhynchos laysanensis* and *A. p. wyvilliana*, both from Hawaii). Habitat destruction is stated to be a threat to 15 of the species and subspecies. Introduced species (domestic livestock in the case of the mammals, and cats, rats, dogs and mongooses in the case of the birds) are a further threat to 15 of the species and subspecies. The introduced species prey on the birds, and outcompete the mammals for pasture and water. In addition, domestic livestock are a source of diseases to which several of the wild mammals are vulnerable. The banteng (*Bos javanicus*), the gaur (*B. gaurus*), the American bighorn sheep (*Ovis canadensis*), the wild water buffalo (*Bubalus arnee*), and the Asiatic wild ass (*Equus hemionus*) have had their numbers reduced by diseases spread by livestock.

Aquaculture

The gene pools of many fish species of actual or potential aquaculture value are rapidly being reduced, a number having disappeared already. In Lake Vanern, Sweden, the largest lake in Western Europe, eight important salmon (*Salmo salar*) and brown trout (*S. trutta*) populations have become extinct in the last 80 years or so. All were destroyed by hydro-electric plants, which drowned the spawning areas, or blocked passage to them. These problems also threaten the remaining four populations (two of salmon, two of trout) (150). In Nova Scotia, Canada, nine distinct gene pools of *S. salar* have been exterminated because of acid rain. Because of atmospheric pollution by power stations and industry, precipitation in the region is 10-15 times more acid than normal. As a result, the acidity of nine spawning rivers has risen above the levels that early feeding salmon fry (the most sensitive stage of the life cycle) can survive. For the same reason, another 13 salmon rivers, and hence another 13 gene pools, are close to extinction, and an additional nine are considered borderline (179). In Norway, too, acid rain has left hundreds of lakes without fish and removed the salmon from many river systems (61).

In the United States many unique populations of the cutthroat trout (*S. clarki*) are threatened, or have disappeared already, because of a combination of habitat destruction (logging too close to streams, pollution, water diversion) and the effects of introduced species (competition, predation, and loss of genetic identity through hybridisation) (122). Although native populations of cutthroat trout and rainbow trout (*S. gairdneri*) have evolved isolating mechanisms so that the two species can coexist in the same waters without interbreeding, no such barriers seem to exist when hatchery reared rainbow trout are introduced. Many native populations of cutthroat trout in Montana have consequently been lost through hybridisation with introduced rainbow trout (1). Many other populations in California, Nevada, Wyoming, Utah, Colorado, Idaho and New Mexico are threatened in the same way (122). It has been estimated that virtually all of the original populations of cutthroat trout in the interior of the US have disappeared during the past 100 years (81).

Reductions in the genetic diversity of wild species of value for aquaculture have been observed mostly in freshwater and anadromous (reproducing in fresh waters) species. This is to be expected, since fresh waters are much more confined than the sea and are exposed to far greater changes. It is now likely that overfishing of marine fish stocks is exerting selective pressure on gene pools. Evidence of such effects has come from some of the freshwater fisheries, notably those of Africa's Great Lakes. For example, reproductively mature *Tilapia nilotica* in Lake George, Uganda, are significantly smaller (on average about a third shorter) than they were 20 years ago. Almost certainly this is due to overfishing, which tends to capture the larger individuals leaving only the smaller ones to survive and reproduce (41). This phenomenon, known as negative selection, has also occurred in forestry, when the best trees are logged, leaving regeneration of the stand to the rejects.

6. Conservation of wild genetic resources

There are two ways of conserving wild genetic resources: *in situ* (in nature reserves) and *ex situ* (in zoos and gene banks). Both are needed. Neither is being done as effectively as it should.

In situ conservation is the maintenance of a wild gene pool in its native habitat, in a national park or nature reserve, for example. *Ex situ* conservation is the maintenance of the resource outside its native habitat: either the whole organism (plant or animal) in a botanical garden, plantation, zoo or breed farm; or just the germplasm (seed, pollen, sperm, ova, budwood or cells) in a seed bank, sperm bank, and so on.

The International Board for Plant Genetic Resources (IBPGR) distinguishes three types of *ex situ* crop germplasm collection (130), which can usefully be applied to all genetic resources and to *in situ* as well as *ex situ* systems:

* 1. *Base collections* for long term maintenance of genetic material.
* 2. *Active collections* for medium term maintenance of genetic material, multiplication and distribution, evaluation and documentation.
* 3. *Working collections* serving the short term needs of individual breeders and breeding programmes.

The primary aim of a base collection (type 1) is the security of the germplasm it conserves. The primary aim of a working collection (type 3) is the convenience of the breeders and breeding programmes it serves. So as one moves from 1-3 convenience increases but security decreases. This chapter is about maintaining the long term availability of wild genetic resources, and hence about progress in the establishment of *ex situ* and *in situ* base collections.

By far the greatest amount of progress has been made in the *ex situ* conservation of crop genetic resources. Since the IBPGR was set up in 1974, it has promoted the development of an impressive network of national and international base collections covering more than 20 of the world's most important crops. These include wheat, rice, maize, barley, oats, sorghum, millet, potato, *Phaseolus* beans, pigeon pea, peanut, chickpea, peas, cowpea, winged bean, onion and garlic, bell and chili pepper, tomato, eggplant, cabbages, and sugar beet (91). The scale of this achievement can be realised by comparing the situation in 1975, when only eight institutions in the world had facilities for long term seed storage, with the situation seven years later, when the number had quadrupled to 33 (134).

Although most of these collections include wild species, the bulk of the germplasm they maintain is of the crops themselves. There are two good reasons for this. First, the domesticated material is generally more useful than the wild material. Second, it has generally been more threatened. Much of the impetus to the collection and maintenance of genetic resources during the last two decades has been due to the growing recognition that many valuable landraces of crops and livestock have been irretrievably lost and the remainder are rapidly disappearing.

Landraces (primitive or traditional cultivars and breeds) can be made extinct with a suddenness that befalls only a few of the most endangered species. With most wild species, extinction is a fairly gradual process; only species with an extremely small distribution can be eliminated completely in, say, one episode of habitat destruction. Landraces have been totally lost as a result of simple abandonment by the farmers who grew them. The landraces may be displaced in one or two seasons by more productive modern cultivars, including emergency seed sent in by famine-relief operations. Between 1930 and the mid-1960s the proportion of the Greek wheat crop grown to traditional landraces fell from 80% to 10% (14).

A 1973 report illustrates dramatically the problems of conserving landraces: "When this survey was made, Afghanistan's native wheats seemed safer from genetic erosion than almost any others. Since then, two years of drought, harvest failures and catastrophic famine have drastically changed the picture. Thousands of tonnes of seeds have been imported. Introduced varieties now predominate in many parts of the country. In mountain areas where previously only indigenous varieties have ever grown, introduced varieties have now widely replaced them. Afghanistan is a warning that genetic erosion does not follow a predictable course, and that genetic conservation programmes must never be relaxed, even in regions considered safe from genetic erosion" (13).

This last statement applies equally forcibly to wild genetic resources. The threats to landraces have not diminished, and are not likely to. The need to collect and conserve them remains a priority. But as we have seen in previous chapters, the importance and vulnerability of the wild gene pools are growing rapidly. These now need to be given the level of attention first accorded to landraces a decade ago.

An IBPGR survey of the 20 major collections of wheat in the world (Australia, Canada [2], Czechoslovakia, France, East Germany, West Germany, Hungary, India, Israel, Italy, Japan [3] Netherlands, Turkey, UK, US [2] and USSR) (38) illustrates the gap. Excluding duplicates, the 20 collections together hold some 4,000 samples of *Aegilops* species and more than 40,000 samples of *Triticum* species. The two most commonly cultivated wheats naturally dominate the collections: there are 9,000 samples of macaroni wheat (the *durum* form of *T. turgidum turgidum*), and 24,500 samples of bread wheat (*T. aestivum vulgare*). The survey classifies as under-represented in the world collection any taxon (species, subspecies or variety) that is represented by fewer than 100 samples. Of the 17 taxa that are considered under-represented, nine are domesticated (various primitive *Triticum* taxa), and eight are wild (species of *Aegilops*).

Superficially, therefore, it would appear that the wild genetic resources of wheat are no worse off than the primitive domesticated genetic resources. The wild taxa, however, are much more patchily represented and hence less of their genetic variation is likely to be protected than is that of the cultivated forms. Of the 17 domesticated taxa (one species + 11 subspecies + five varieties), 12 (or 70%) are represented by samples from each country in which the taxa occur. But of the 24 wild taxa (20 *Aegilops* species + one species and three subspecies of *Triticum*), only five (or 21%) are represented by samples from all of the countries of their natural distribution. There is not necessarily any relationship between the number of countries in which a species occurs and its genetic variation. But the variability of a species cannot be known until it has been adequately sampled; and it is clear from the survey that most of the wild relatives of wheat have not been.

The situation for wheat is good compared with rice and maize. There are 38 active and base collections of rice listed in the *IBPGR directory of germplasm collections* (172): six in developed countries, the rest in developing countries. Only 12 report holdings of wild species, and only five of these (in Japan, Nigeria, the Philippines, the USSR and the US) have long term storage facilities. Of 53 active and base collections of maize (22 in developed countries, 31 in developing countries) (7), only six include wild species and only one of these (in the US) can maintain them long term.

Many more of the germplasm collections of food legumes include wild species: 29 out of 66 (31 in developed countries; 35 in developing countries). Ten of these are base collections: six among developed countries — Belgium, German Democratic Republic, Italy, Sweden, the US and the USSR; and four among developing countries: Argentina, Colombia, Costa Rica and Nigeria (84). However, these collections usually include a greater number of crops (*Phaseolus* beans, *Vigna* beans, soybeans, peanuts, chickpeas, lentils etc) than the cereal collections, and there are much fewer individual accessions (samples). For example, the US collection of soybean and its wild relatives, the most comprehensive in the world, contains some 7,200 accessions. This is quite small compared with the collections of cereal crops (such as wheat, rice and sorghum) which are from two to five times larger (84). Even so, soybean cultivars make up 90% of the collection; the wild GP1 species *Glycine soja* accounts for only 7% and samples of the six GP3 species for only 3%, a mere 197 samples (85).

Seed-propagated crops with seeds that can be dried and cold-stored for up to a century, like the cereals and legumes, are relatively easy to maintain. Crops that are maintained vegetatively (like banana), or crops like cacao whose seeds cannot be dried and then stored without killing them, are harder to keep safe. There are many crops, including timber trees, that fall into this category: mango, rubber, chestnuts, oaks, walnuts, cinnamon, avocado, mahoganies, coffee, citrus and cacao are among them (101). With such crops there are severe physical contraints on the amount, and therefore the diversity, of the germplasm that can be stored.

There are only five major banana collections in the world, in Honduras, Jamaica, the Philippines, Papua New Guinea and India. Their combined holdings consist of 1,366 cultivars and 144 samples of wild species. An IBPGR group of experts on the genetic resources of bananas and plantains agreed that while the collections contained a fair representation of the cultivars, the holdings of wild forms were small (86). Another group of experts discussed the conservation of cacao germplasm and concluded that the major collections (in Trinidad, Brazil, Ecuador, Costa Rica and Venezuela) "are not representative of the gene pool of cocoa and collecting is urgent in many regions" (89).

The IBPGR experts on coffee have drawn attention to the relative absence of wild and semi-wild *Coffea arabica* in collections, the poor representation of *C. canephora* and *C. liberica*, and the almost total lack of coverage of wild coffee species from Madagascar and neighbouring islands: species that may have potential in breeding since they yield beans that are very low in caffeine (88). Another IBPGR working group, on the genetic resources of the sweet potato, has emphasised that cultivated material held by the 19 collections that have reported on their holdings does not appear to be very diverse. The group also expressed "great concern" at the even smaller holdings of related wild species (90).

The *ex situ* collections of crop genetic resource centres and provenance collections of tree breeders are probably doing quite well compared to other *ex situ* systems such as zoos and botanical gardens. Within-species variation is after all the stock-in-trade of agricultural and silvicultural gene banks. Zoos and botanical gardens tend to operate at the species level, since their logistical problems are even greater than those of cacao or banana or sugarcane conservationists. William G Conway, director of the New York Zoological Society, has remarked that all the zoo animal enclosures in the world could comfortably fit within the borough of Brooklyn (34). It is therefore sensible for them to concentrate their captive breeding efforts towards providing a safety net for species close to extinction, rather than attempting to maintain a wide range of the gene pools of less endangered species. This would include saving the wild relatives of livestock species from extinction, as captive breeding programmes have very successfully done in the case of the European bison (*Bison bonasus*) and Przewalski's horse (*Equus ferus*). The former became extinct in the wild, but was increased from surviving zoo animals and has been reintroduced in Poland. The latter is believed to be extinct in the wild and exists only as a zoo animal. The present population of about 265 horses is descended from nine animals obtained from Mongolia at the beginning of this century (12; 35). According to Conway, the population today could have been as high as 1,000 but for the application of arbitrary restrictions in the breeding programmes, limiting the male line to a few stallions. A substantial portion of the population, he writes, "is beginning to show the effects of inbreeding depression in lowered vitality and various anomalies" (35).

Botanical gardens face lesser but similar problems. Like zoos they can and do combine whole organism maintenance and germplasm maintenance. But since crop relatives are generally smaller than livestock relatives and many plants are inbreeders, whereas all the animals are outbreeders, botanical gardens are less limited than zoos in the role they can play. Similarly, the maintenance of plant germplasm (pollen, seeds, etc), while not without its problems (some of which have yet to be overcome), is generally easier than the maintenance of animal germplasm (semen, ova etc). So one can envisage botanical gardens making a greater contribution to *ex situ* conservation of wild genetic resources. For example, they could maintain samples of a wide range of gene pools of locally occurring crop relatives; and could act as bridging institutions between agricultural and silvicultural *ex situ* gene banks on one side, and *in situ* gene banks on the other.

In situ gene banks

At present, however, *in situ* gene banks are more a hope than a reality. There are many national parks, nature reserves, biosphere reserves and countless other protected areas, but as far as we are aware the number of genuine *in situ* gene banks can be counted on the fingers of one hand. The USSR has apparently established a reserve in the Kopet-Dag mountains (Turkmen SSR) to protect wild forage grasses, wild apricot, pistachio and almond, and one in the Caucasus to protect wild wheat and wild fruit trees (18). Sri Lanka has set up a reserve for wild medicinal plants, and India is planning sanctuaries to protect wild relatives of banana, citrus, rice, sugarcane and mango (87).

Otherwise, the world's protected areas continue to be designed and operated with little if any consideration for the maintenance of wild genetic resources. In 1980, on behalf of IBPGR and the IUCN, we surveyed the government agencies responsible for protected areas in a sample of 50 countries. The survey was concerned only with the wild relatives of crops and not with other wild genetic resources. Nevertheless, from the 30% response it was clear that by and large protected areas are currently ill-equipped to service the potential users of the genetic resources they may maintain.

A minimum requirement is documentation. Yet fewer than half of the countries which replied had compiled lists of the species occurring in their protected areas — and then only for a small minority of the reserves. For the time being, therefore, it is impossible to tell how many wild relatives of crops occur in protected areas, and consequently it is impossible to decide whether new areas are needed and, if so, where. Lack of information also means that it is seldom possible to determine whether a particular wild gene pool requires special management. It may be that a reserve is ostensibly protecting populations of wild cereals or wild grapes, but that in practice those populations are being overgrazed by herbivores that the reserve is also attempting to maintain.

In addition, the absence of documentation reduces the value of nature reserves to the users of genetic resources; even if a reserve is protecting valuable gene pools, there is no way the users can know about it. Users are likely to face other obstacles as well. Fewer than 15% of the countries surveyed permit the collection of germplasm (seeds, budwood etc), and almost 25% do not allow collection for any purpose.

Another unmet need is adequate liaison between the agencies responsible for protected areas, and those responsible for research and the protection of crop genetic resources. Even if important wild genetic resources *are* being maintained in a protected area, the potential users may not know about it, may not be allowed access to the resources, or may be hampered by the lack of facilities for research or standby storage of any material collected.

In the late 1960s and early 1970s, before the establishment of IBPGR and the major international *ex situ* gene banks, and before the establishment of the first biosphere reserves, the scientific leaders of the genetic resources community were emphatic about the need for both *ex situ* and *in situ* conservation. Recommendation 39 of the United Nations Conference on the Human Environment (Stockholm, 1972), for example, recommended both static (= *ex situ*) and dynamic (= *in situ*) ways of preserving wild plant and animal gene pools.

Implicit (and sometimes explicit) in the concept of biosphere reserves, as it was developed in the early 1970s, was the idea of conserving genetic diversity at several levels: the ecosystem, the species and the gene pool. But in practice, biosphere reserves, like national parks and other protected areas, have been selected and managed with the first two levels of diversity in mind, not the third.

Why is it that the *in situ* conservation of wild genetic resources has made so little progress? In our analysis, the reason is a combination of sectoralism and prejudice. The various kinds of protected areas have been set up with numerous, sometimes conflicting, aims:

* to protect wildernesses and sacred groves and other areas symbolic of the relationship between a particular society and the natural world;
* to provide amenity and recreation;
* to serve as sites for education and scientific research;
* to safeguard ecosystems that play key positions in ecological processes, such as the role of watershed forests in regulating local and regional hydrology;
* to act as a reserve in the regeneration of a resource such as timber, game, or sport fish;

* to protect habitats critical for the survival of an endangered species;
* to conserve entire ecosystems that may be threatened, such as a coral reef.

Agencies administering protected areas have tended to emphasise some types of goal at the expense of others, depending on their mandates: water management, fisheries, forestry, national parks, wildlife, tourism and recreation, science etc.

There have been only two overarching, integrating approaches to the jungle of conflicting mandates: the philosophy of *national parks*, promoting a combination of preservation of natural areas and biological diversity with facilities for public recreation; and the concept of a world network of areas conserving representative and unique examples of the biological diversity of the planet, of which *biosphere reserves* coordinated by UNESCO (United Nations Educational, Scientific & Cultural Organization) are the most coherent expression.

There is a place for *in situ* gene banks in both schemes — which overlap, of course. However, the criteria for selecting protected areas under these schemes have concentrated on features of outstanding beauty and interest, and examples of the diversity of ecosystems and of species, rather than on variation *within* species. Thus Dr Michel Batisse (deputy assistant director-general for science at UNESCO) explains (10) that "each biosphere reserve should include one or more of the following:

(i) representative examples of natural biomes;
(ii) unique communities or areas with unusual features of exceptional interest;
(iii) examples of harmonious landscape resulting from traditional patterns of land-use; and/or
(iv) examples of modified or degraded ecosystems that are capable of being restored to more-or-less natural conditions."

There is nothing in this that would prevent a biosphere reserve being an *in situ* gene bank (or vice versa), and in fact the biosphere reserve system has great potential for the development of *in situ* gene banks. But nor is there anything that makes clear that within-species variation is one of the objects of conservation.

Indeed, Dr Batisse wrote earlier that one of the primary objectives of biosphere reserves is to safeguard "the genetic diversity on which their continuing evolution depends" (10). The phrase "on which their continuing evolution depends" reveals an unconscious, and traditional, bias. There are two reasons for conserving within-species variation. First, because genetic variation is essential for species to adapt and survive. Second, because genetic variation is the raw material of domestication and of the continued survival and improvement of domesticates. Both reasons are equally valid, but each requires distinctly different approaches to conservation. Conservationists have generally thought only of the first rather than of the second.

The difficulties

While wildlife conservationists and the protected areas community have been slow to put genetic resources on their agenda, the genetic resources community has been just as slow in turning its attention to *in situ* conservation. For reasons already given, *ex situ* conservation took priority, but in addition many people in the genetic resources community have been frankly sceptical that *in situ* conservation is either necessary or possible. They see three main obstacles that rule out protected areas as a means of conserving wild genetic resources: problems of use, problems of security, and problems of coverage.

First, there are problems of use. *Ex situ* gene banks can be sited wherever is most convenient to the users of the resources they house. *In situ* gene banks by definition have to be where the resources are. Wild species protected in nature reserves will often be far from the breeders that need them — in some cases continents apart. Examples are bananas (wild in Southeast Asia, cultivated in the Caribbean and Central America), rubber (wild in South America, cultivated in Southeast Asia), and eucalyptus (wild in Australia but grown all over the world). This is a valid objection only if *in situ* gene banks are intended as active or working collections, which clearly they cannot be except in special circumstances. One such circumstance would be reserves protecting the genetic resources of a country's timber trees, which will often be close to (and may include) areas being logged and grown to plantations. Generally, however, the main contribution of *in situ* gene banks will be as base collections, safeguarding wild genetic resources over the long term. The location of a base collection does not matter as long as it is safe.

Second, there are problems of security. Genetic resource users fear that *in situ* gene banks would be vulnerable to periodic disturbance by poachers, firewood collectors and livestock, or to outright destruction by development, or by the pressures of population growth. They could be invaded by farmers, or be dismembered by the flick of a legislative pen or the proffer of a bribe. *Ex situ* gene banks, it is argued, can keep everything safely under one roof and are much easier to protect. This is true. Protected areas are intrinsically more vulnerable than *ex situ* collections to changes in land use for whatever cause. However, it is easy to exaggerate both the insecurity of *in situ* gene banks and the security of *ex situ* gene banks. The greater the perceived value of a protected area the longer it is likely to last, and conserving wild genetic resources will increase a protected area's value to society.

For their part, *ex situ* collections are vulnerable to human error. Many have been lost in the past and it is unlikely that future losses, while they can be reduced, can be avoided altogether.

* From 1929 to 1931, two US Department of Agriculture seed explorers collected 4,000 samples of soybean seed from China, Japan and Korea; today less than a third of their collection still survives, the rest having been thrown out or lost (83).

* Maintenance problems have led to considerable losses of sugarcane clones (91). Out of some 1,000 clones collected from Papua New Guinea from 1875 to 1955, only 281 were remaining in 1958, and 204 by 1975. A major sugarcane collection which included important wild material was reduced by half in the 18 years since the collection was made (15).
* Several valuable collections of wild grape species have been almost entirely lost due to lack of continuity in the breeding programmes that initiated the collections; remaining collections are experiencing problems of maintenance (137).
* More sensitive species like wild cocoa suffer maintenance problems right from the point of collection, well before they even arrive at the *ex situ* gene bank: collections in Brazil have been lost due to airline delays (137).
* The difficulties of maintaining wild species *ex situ* arise either because they are poorly adapted to the location, such as the wild African rices *Oryza barthii* and *O. longistaminata* at the International Rice Research Institute in the Philippines (69); or because they are difficult to germinate and grow out, such as some of the wild sunflowers (138) and chickpeas (117).

The third problem with *in situ* gene banks is that of coverage. Attempts to safeguard adequate samples of the genetic variation of wild species may, it is argued, take up so much land that they are quite impractical. This may or may not be so. In practice, the conservation of wild genetic resources has been so badly neglected that little is known of the distribution of potentially valuable wild gene pools. It is very doubtful that all the important wild germplasm can be protected in any way — either *ex situ* or *in situ*. It is already disappearing. But it seems highly likely that a worthwhile proportion of surviving genetic variation could be saved, and that the establishment of effective *in situ* gene banks would increase that proportion.

One advantage of *in situ* gene banks is that they can serve several sectors at once. Gene pools of value to agriculture, forestry and aquaculture may very well overlap and hence be conserved in the same protected area. Another advantage is that evolution can continue within a protected area. This is especially important for pest and disease resistance. In the wild, resistant species can co-evolve with parasites and pathogens, providing the breeder with a dynamic reservoir of resistance that is lost when the material is transferred to the deep freeze of an *ex situ* gene bank. Areas that have a rich diversity of pest and disease races, and a corresponding richness in the diversity of resistant plants, would be excellent candidates as *in situ* gene banks: parasite and pathogen parks. Central Mexico (late blight and wild potatoes), central US (*Phylloxera* and wild grapes), Israel (rusts and wild oats), southwestern USSR (rusts and wild wheats) might be good places to start.

Yet another advantage of *in situ* gene banks is that they can double as living laboratories. Maintenance of a species in its natural habitat allows the breeder to study its ecology and so obtain information that might otherwise be overlooked. Several valuable characteristics of wild tomatoes have been discovered in this way: high soluble solids, tolerance of intense tropical moisture and temperatures, tolerance of saline soils, insect resistance and drought resistance (143; 146).

The problem with discussing the pros and cons of *ex situ* and *in situ* gene banks is that it appears that they are in competition. They are not. As we stated at the beginning of this chapter, both are needed. They complement each other, the strengths of one compensating for the other's weaknesses. Fortunately, a convergence of opinion appears to be emerging among genetic resource users and the wildlife conservationists and protected areas community. IBPGR took the initiative in commissioning a position report on the *in situ* conservation of the wild relatives of crops. And the United Nations Environment Programme (UNEP) and FAO have been active in getting experts together to decide on the best ways of getting *in situ* conservation off the ground. These meetings have included one on the *in situ* conservation of forest genetic resources (51) and one on conserving the genetic resources of fish (47).

IUCN, the leading scientific nongovernmental conservation body, is actively pursuing ways of integrating the new concept of *in situ* gene banks into the well-established procedures of conserving threatened species and managing national parks and nature reserves. UNESCO, too, has recognised that biosphere reserves and *in situ* gene banks are perfect partners.

Some wild gene pools can be maintained passively: leave them alone, protect them from harm, and they will persist. Other wild gene pools will have to be maintained actively: many wild species in primary gene pools flourish only in fairly disturbed environments, and the *in situ* gene banker will need to maintain that level of disturbance to maintain the populations concerned. The concept of core and buffer zones in the biosphere reserves provides very well for the combination of preservation and active management that much *in situ* gene banking is likely to call for. New ways of maintaining, and new ways of using the newest resource are being developed, as they must be, side by side.

REFERENCES

0a. Adansi, MA. 27 April 1982, personal communication.
1. Allendorf, FW & SR Phelps. 1981. Isozymes and the preservation of genetic variation in salmonid fishes. In: N Ryman (editor) *Fish gene pools*: 37–52.
2. Alston, FH. 1977. Practical aspects of breeding for mildew (*Podosphaera leucotricha*) resistance in apples. *Proceedings of Eucarpia Fruit Section. Symposium VII.* 7–10 September 1976. Wageningen: 4–17.
2a. Alston, FH. 24 February 1981, personal communication.
3. Amerine, MA & VL Singleton. 1976. *Wine: an introduction.* 2nd edition. University of California Press. Berkeley, Los Angeles, London. 370 pp.
3a. Angkapradipta, P. 23 July 1981, personal communication.
4. Anonymous, 1980. *North Carolina State University – industry cooperative tree improvement program 24th annual report.* North Carolina State University. School of Forest Resources. Raleigh. 63 pp.
5. Arasu, NT & N Rajanaidu. 1975. Conservation and utilization of genetic resources in the oil palm. In: JT Williams, CH Lamoureux & N Wulijarni-Soetjipto (editors). *South East Asian plant genetic resources*: 182–186.
6. Ayad, G & NM Anishetty. 1980. *Directory of germplasm collections. I. Food legumes.* IBPGR. AGP:IBPGR/80/45. Rome. 22 pp.
7. Ayad, G, J Toll & JT Esquinas-Alcazar. 1980. *Directory of germplasm collections. III. Cereals. 2. Maize.* IBPGR. AGP:IBPGR/80/90. Rome. 23 pp.
8. Bardach, JE, JH Ryther & WO McLarney. 1972. *Aquaculture: the farming and husbandry of freshwater and marine organisms.* John Wiley & Sons, New York, Chichester, Brisbane, Toronto. 868 pp.
9. Barnard, C. 1972. *Register of Australian herbage plant cultivars.* CSIRO. Division of Plant Industry. Canberra, 260 pp.
10. Batisse, M. 1982. The biosphere reserve: a tool for environmental conservation and management. *Environmental Conservation* &: 101–112.
11. Baum, BR. 1977. *Oats: wild and cultivated. A monograph of the genus Avena L. (Poaceae).* Agriculture Canada. 463 pp.
11a. Bedigian, D. November 1981, personal communication.
12. Benirschke, K, B Lasley & O Ryder. 1980. The technology of captive propagation. In: ME Soule & BA Wilcox (editors). *Conservation biology: an evolutionary–ecological perspective.* 225–242.

13. Bennett, E. 1973. Near East: cereals: Afghanistan. In: OH Frankel (editor). *Survey of crop genetic resources in their centres of diversity: first report*: 22-24.
14. Bennett, E. 1973. Mediterranean: wheats of the Mediterranean basin. In: OH Frankel (editor). *Survey of crop genetic resources in their centres of diversity: first report:* 1-8.
15. Berding, N & H Koike. 1980. Germplasm conservation of the *Saccharum* complex: a collection from the Indonesian archipelago. *Hawaiian Planters' Record* 59: 87-176.
16. Bernsten, RH, BH Siwi & HM Beachell. 1981. *The development and diffusion of rice varieties in Indonesia*. Paper presented at the International Rice Research Conference. 29 April 1982. 69 pp.
16a. Blaak, Ir G. 19 October 1981, personal communication.
17. Bogdan, AV. 1977. *Tropical pasture and fodder plants (grasses and legumes)*. Longman, London, New York. 475 pp.
18. Brezhnev, DD. 1975. Plant exploration in the USSR. In: OH Frankel & JG Hawkes (editors). *Crop genetic resources for today and tomorrow*: 147-150.
19. Brooke, CH & ML Ryder. 1978. *Declining breeds of Mediterranean sheep*. FAO Animal Production and Health Paper 8. FAO. Rome.
20. Brown, AG. 1975. Apples. In: J Janick & JN Moore (editors). *Advances in fruit breeding*: 3-37.
21. Budin, K. 1973. The use of wild species and primitive forms in agricultural crop breeding in the USSR. In: JG Hawkes & W Lange (editors). *Proceedings of a conference on European and regional gene banks*: 87-97.
22. Burley, J. 1976. Genetic systems and genetic conservation of tropical pines. In: J Burley & BT Styles (editors). *Tropical trees: variation, breeding and conservation.*: 85-100.
23. Burley, J & BT Styles (editors). 1976. *Tropical trees: variation, breeding and conservation*. Linnean Society Symposium Series Number 2. Academic Press. London, New York. 243 pp.
24. Carlisle, A & AH Teich. 1971. *The costs and benefits of tree improvement programs*. Canadian Forestry Serivce Publication No 1302. Ottawa. 5 pp.
25. Carter, JF (editor). 1978. *Sunflower science and technology*. American Society of Agronomy, Crop Science Society of America, Soil Science Society of America, Inc. Madison. 505 pp.
26. Cauquil, J & CD Ranney. 1967. *Studies on internal infection of green cotton bolls and the possibilities of genetic selection to reduce boll rot*. Mississippi Agricultural Experiment Station. Technical Bulletin 53. 24 pp.
27. Chalmers, WS. 1972. *The conservation of wild cacao populations: the plant breeders' most urgent task*, 4th International Cocoa Conference. January 1972. Trinidad and Tobago.
28. Chang, TT. 1975. Exploration and survey in rice. In: OH Frankel & JG Hawkes (editors). *Crop genetic resources for today and tomorrow*: 159-165.

29. Chang, TT. 1976. Rice. In: NW Simmonds (editor). *Evolution of crop plants*: 98–104.
30. Chippendale, GM & L Wolf. 1981. *The natural distribution of Eucalyptus in Australia*. Australian National Parks and Wildlife Service. Special Publication 6. Canberra. 192 pp.
31. Clements, RJ & RJ Williams. 1980. Genetic diversity in *Centrosema*. In: RJ Summerfield & AH Bunting (editors). *Advances in legume science*: 559–567.
32. Cocks, PS & JR Phillips. 1979. Evolution of subterranean clover in South Australia. I. The strains and their distribution. *Australian Journal of Agricultural Research* 30: 1035–1052.
33. Cocks, PS, MJ Mathison & EJ Crawford. 1980. From wild plants to pasture cultivars: annual medics and subterranean clover in southern Australia. In: RJ Summerfield & AH Bunting (editors). *Advances in legume science*: 569–596.
33a. Cocks, PS. October 1981, personal communication.
33b. Cocks, PS. December 1981, personal communication.
34. Conway, WG. No date. *Gene banks for higher animals*.
35. Conway, WG. 1980. An overview of captive propagation. In: ME Soule & BA Wilcox (editors). *Conservation biology: an evolutionary–ecological perspective*: 199–208.
36. Corbet, GB & JE Hill. 1980. *A world list of mammalian species*. British Museum (Natural History). London. Cornell University Press. Ithaca. 226 pp.
37. Correll, DS. 1962. *The potato and its wild relatives*. Texas Research Foundation. Renner (Texas). 606 pp.
38. Croston, RP & JT Williams. 1981. *A world survey of wheat genetic resources*. IBPGR. AGP:IBPGR/80/59. Rome.
39. Cummins, JN & HS Aldwinckle. 1979. Breeding tree crops. In: MK Harris (editor). *Biology and breeding for resistance to arthropods and pathogens in agricultural plants*: 528–545.
40. Damania, AB & JT Williams. 1980. *Directory of germplasm collections. II. Root crops*. IBPGR. Rome. 54 pp.
41. Daniels, J, P Smith & N. Paton. 1975. The origin of sugarcanes and centres of genetic diversity in *Saccharum*. In: JT Williams, CH Lamoureux & N Wulijarni-Soetjipto (editors). *South East Asian plant gentic resources*: 91–107.
42. Doebley, JF & HH Iltis. 1980. Taxonomy of *Zea* (Gramineae). I. A subgeneric classification with key to taxa. *American Journal of Botany* 67: 982–993.
43. Durbin, RD (editor). 1979. *Nicotiana: procedures for experimental use.* USDA. Technical Bulletin 1586. 124 pp.
44. FAO. 1975. *Pilot study on conservation of animal genetic resources*. FAO.

Rome. 60 pp.
45. FAO. 1980. *In situ conservation of forest genetic resources: secretariat note.* FAO. Rome.
46. FAO. 1981. *1980 FAO production yearbook.* FAO Statistics Series No 34. Rome. 296 pp.
47. FAO. 1981. *Conservation of the gentic resources of fish: problems and recommendations.* Reports of the expert consultation on the genetic resources of fish. 9–13 June 1980. Rome. Fao Fisheries Technical Paper No 217. FIRI/T217. Rome. 43 pp.
48. FAO. 1981. *Forest genetic resources information – no 10.* FAO. Rome. 55 pp.
49. FAO. 1981. *Data book on endangered forest tree species and provenances.* FO:MISC/81/11. Rome. 64 pp.
50. FAO. 1981. *Report of the FAO/UNEP/IBPGR international conference on crop genetic resources.* 6–10 April 1981. Rome. FAO. AGP:1981/M/6. Rome. 80 pp.
51. FAO & UNEP. 1981. *Report on the FAO/UNEP expert consultation on in situ conservation of forest genetic resources.* 2–4 December 1980. Rome. 34 pp.
52. Feldman, M & ER Sears. 1981. The wild gene resources of wheat. *Scientific American* 244: 98–109.
52a. Feldman, M. 29 July 1981, personal communication.
52b. Fernie, LM. 21 July 1981, personal communication.
53. Ferwerda, FP. 1976. Coffees. In: NW Simmonds (editor). *Evolution of crop plants*: 257–260.
54. Fowler, DP. 1978. Population improvement and hybridization. *Unasylva* 30: 21–26.
55. Frankel, OH. 1973. Africa: coffee (introduction). In: OH Frankel (editor). *Survey of crop genetic resources in their centres of diversity: first report*: 69.
56. Frankel, OH (editor). 1973. *Survey of crop genetic resources in their centres of diversity: first report.* FAO, IBPGR & ICSU. 164 pp.
57. Frankel, OH & E Bennett (editors). 1970. *Genetic resources in plants – their exploration and conservation.* International Biological Programme Handbook No 11. Blackwell Scientific Publications. Oxford, Edinburgh. 554 pp.
58. Frankel, OH & JG Hawkes (editors). 1975. *Crop genetic resources for today and tomorrow.* International Biological Programme 2. Cambridge University Press, Cambridge, London, New York, Melbourne, 492 pp.
59. Frankel, OH & ME Soule. 1981. *Conservation and evolution.* Cambridge University Press. Cambridge, London, New York, New Rochelle, Melbourne, Sydney. 327 pp.
59a. Freytag, GF. 6 October 1981, personal communication.
60. Galet, P. 1979. *A practical ampelography: grapevine identification.* Translated and adapted by LT Morton, Cornell University Press, Ithaca, London. 248 pp.

61. Gjedrem, T. 1981. Conservation of fish populations in Norway. In: N Ryman (editor). *Fish gene pools*: 33–36.
61a. Glendinning, DR. 14 July 1981, personal communication.
61b. Gladstones, JS. 29 September 1981, personal communication.
62. Golodriga, PY & IA Souyatinou. 1981. *Vitis amurensis*. Habitat. Aptitudes culturales et technologiques. Variabilité de l'espèce. *Bulletin de l'OIV*. 54: 971–982.
63. Hahn, SK, AK Howland & ER Terry. 1973. Cassava breeding at IITA. *Proceedings of the 3rd Symposium of the International Society for Tropical Root Crops*. 2–9 December 1973: 4–10.
64. Hahn, SK, AK Howland & ER Terry. 1980. Correlated resistance of cassave to mosaic and bacterial blight diseases. *Euphytica* 29: 305–311.
65. Hanh, SK, ER Terry & K Leuschner, 1980. Breeding cassava for resistance to cassava mosaic disease. *Euphytica* 29: 673–683.
65a. Hahn, SK. 1 September 1981, personal communication.
66. Harlan, JR. 1971. Agricultural origins: centers and noncenters. *Science* 174: 468–474.
67. Harlan, JR. 1973. Genetic resources of some major field crops in Africa. In: OH Frankel (editor). *Survey of crop genetic resources in their centres of diversity: first report*: 45–64.
68. Harlan, JR. 1975. *Crops and man*. American Society of Agronomy. Madison (Wisconsin). 295 pp.
69. Harlan, JR. 1981. *Evaluation of wild relatives of crop plants*. Paper presented at: FAO/UNEP/IBPGR Technical Conference on Crop Genetic Resources. 6–10 April 1981. Rome.
70. Harlan, JR & D Zohary. 1966. Distribution of wild wheats and barleys. *Science* 153: 1074–1080.
71. Harlan, JR & JMJ de Wet. 1971. Toward a rational classification of cultivated plants. *Taxon* 20: 509–517.
72. Harland, SC. 1970. Gene pools in the new world tetraploid cottons. In: OH Frankel & E. Bennett (editors). *Genetic resources in plants – their exploration and conservation*: 335–340.
73. Harris, MK (editor). 1979. *Biology and breedings for resistance to arthropods and pathogens in agricultural plants*. Proceedings of a short course entitled "International short course in host plant resistance". 22 July–4 August 1979. College Station, Texas.
74. Harvey, PH, CS Levings III & EA Wernsman. 1972. The role of extrachromosomal inheritance in plant breeding. *Advances in Agronomy* 24: 1–27.
75. Hawkes, JG. 1945. The indigenous American potatoes and their value in plant breeding. 1. Resistance to disease. 2. Physiological properties, chemical composition and breeding capabilities. *Empire Journal of Experimental Agriculture* 13: 11–40.

76. Hawkes, JG, 1979. Genetic poverty of the potato in Europe. In: AC Zeven & AM van Harten (editors). *Proceedings of the conference broadening the genetic base of crops*: 19–27.
76a. Hawkes, JG. 8 April 1981, personal communication.
77. Hawkes, JG & W Lange (editors). 1973. *Proceedings of a conference on European and regional gene banks*. 10–15 April 1972. Izmir (Turkey). Eucarpia. Wageningen (Netherlands). 107 pp.
78. Hawtin, GC. 1979. Strategies for the genetic improvement of lentils, broad beans, and chick peas, with special emphasis on research at ICARDA. In: GC Hawtin & GJ Chancellor (editors). *Food legume improvement and development*: 147–154.
79. Hawtin, GC & GJ Chancellor (editors). 1979. *Food legume improvement and development*. Proceedings of a workshop held at the University of Aleppo, Syria. 2–7 May 1978. ICARDA & IDRC. IDRC-126e. Ottawa. 216 pp.
80. Hedegart, T. 1976. Breeding systems, variation and genetic improvement of teak (*Tectona grandis* L.f.). In: J Burley & T Styles (editors). *Tropical trees: variation, breeding and conservation*: 109–123.
80a. Henzell, EF. 28 August 1981, personal communication.
81. Hodgson, RE (editor). 1961. *Germ plasm resources*. A symposium presented at the Chicago meeting of the American Association for the Advancement of Science. 28–31 December 1959. AAAS. Washington, DC.
83. Hymowitz, T. 1976. Soyabeans. In: NW Simmonds (editor). *Evolution of crop plants*: 159–162.
84. Hymowitz, T & CA Newell. 1977. Current thoughts on origins, present status and future of soyabeans. In: DS Seigler (editor). *Crop resources*: 197–209.
85. Hymowitz, T & CA Newell. 1980. Taxonomy, speciation, domestication, dissemination, germplasm resources and variation in the genus *Glycine*. In: RJ Summerfield & AH Bunting (editors). *Advances in legume science*: 251–264.
86. IBPGR. 1978. *Report of IBPGR working group on the genetic resources of bananas and plantains*. 5–7 July 1977. Rome.
87. IBPGR. 1978. *Report of IBPGR workshop on South Asian plant genetic resources*. IBPGR Secretariat. Rome.
88. IBPGR. 1980. *Report of a meeting of a working group on coffee*. 11–13 December 1979. Rome.
89. IBPGR. 1981. *IBPGR working group on genetic resources of cocoa*. IBPGR. AGP:IBPGR/80/56. Rome. 25 pp.
90. IBPGR. 1981. *IBPGR working group on the genetic resources of sweet potato*. 5–7 August 1980. Charleston (South Carolina, USA). IBPGR. AGP:IBPGR/80/63. Rome. 30 pp.
91. IBPGR. 1982. *IBPGR consultative group on international agricultural*

101

   research annual report 1981. IBPGR. Rome. 119 pp.
92. Iizuka, M. 1975. Wild species of *Pyrus* and *Vitis* in Japan. In: T. Matsuo (editor). *Gene conservation – exploration, collection, preservation and utilization of genetic resources*: 25–32.
93. Iltis, HH & JF Doebley. 1980. Taxonomy of *Zea* (Gramineae). II. Subspecific categories in the *Zea mays* complex and a generic synopsis. *American Journal of Botany* 67: 994–1004.
93a. Iltis, HH. 14 July 1981, personal communication.
94. Imle, EP. 1978. *Hevea* rubber—past and future. *Economic Botany* 32: 264–277.
95. Jain, SK, AM Olivieri & J Fernandez-Martinez. 1977. Serpentine sunflower, *Helianthus exilis*, as a genetic resource. *Crop Science* 17: 477–479.
96. Janick, J & JN Moore (editors). 1975. *Advances in fruit breeding*. Purdue University Press. West Lafayette (Indiana). 622 pp.
97. Jewell, PA, C Miller & JM Boyd (editors). 1974. *Island survivors: the ecology of the soay sheep of St Kilda*. The Athlone Press. London. 386 pp.
98. Jolly, MS. 1980. Distribution and differentiation in *Antheraea* species (Saturniidae: Lepidoptera). *Paper presented in the XVI International Congress of Entomology*. 3–9 August 1980. Kyoto, Japan. 18 pp.
99. Jolly, MS, SK Sen & MG Das. 1976. Silk from the forest. *Unasylva* 28(114): 20–23.
99a. Kerr, EA. 14 September 1981, personal communication.
100. Khush, GS, KC Ling, RC Aquino & VM Aguiero. 1977. Breeding for resistance to grassy stunt in rice. *Plant Breeding Papers* 1(4b): 3–9.
101. King, MW & EH Roberts, 1979. *The storage of recalcitrant seeds – achievements and possible approaches*. IBPGR. Rome. 96 pp.
102. King, WB. 1979. *IUCN red data book*. Volume 2: Aves. IUCN. Morges (Switzerland).
103. Knott, D & J Dvorak. 1976. Alien germplasm as a source of resistance to disease. *Annual Review of Phytopathology* 14: 211–235.
103a. Laborde, JA. 19 October 1981, personal communication.
104. Lacaze, JF. 1978. Advances in species and provenance selection. *Unasylva* 30: 17–20.
105. Lamb, RC & H Aldwinckle. 1981. Disease resistance in fruit crops. *Pennsylvania Fruit News 1981 Proceedings* 60: 55–59.
106. Lauvergne, JJ. 1975. Disappearing cattle breeds in Europe and the Mediterranean basin. In: FAO. *Pilot study on conservation of animal genetic resources*: 21–41.
107. Leigh, J, J Briggs & W Hartley. 1981. *Rare or threatened Australian plants*. Australian National Parks and Wildlife Service. Special Publication 7. Canberra. 178 pp.
108. Lemeshev, NK. 1973. the sources of wilt-resistance in the world cotton collection. *OEPP/EPPO Bulletin*. 3: 101–105.

109. Leon, J. 1973. Primitive genetic resources in Latin America. In: OH Frankel (editor). *Survey of crop genetic resources in their centres of diversity: first report*: 71–75.
110. Lesins, KA & I Lesins. 1979. *Genus Medicago (Leguminosae): a taxogenetic study*. Dr W Junk Publishers. The Hague, Boston, London. 228 pp.
110a. Ling, KC. 19 January 1982, personal communication.
111. Loegering, WQ, CO Johnston & JW Hendrix, 1967. Wheat rusts. In: KS Quinsenberry & LP Reitz (editors). *Wheat and wheat improvements*: 307–335.
112. Lucas, G & H Synge. 1978. *The IUCN plant red data book*. IUCN. Gland. 540 pp.
113. Lukefahr, MJ & C Rhyne. 1960. Effects of nectariless cottons on populations of three lepidopterous insects. *Journal of Economic Entomology* 53: 242–244.
114. Lukefahr, MJ, CB Cowan, TR Pfrimmer & LW Noble. 1966. Resistance of experimental cotton strain 1514 to the bollworm and cotton fleahopper. *Journal of Economic Entomology* 59: 393–395.
115. Mackay, JHE. 1982. *Register of Australian herbage plant cultivars: supplement to the 1972 edition*. CSIRO. Division of Plant Industry. Melbourne (Australia). 122 pp.
116. Maesen, LJG van der. 1972. *Cicer L., a monograph on the genus with special reference to the chickpea (Cicer arietinum L.) its ecology and cultivation*. Medelingen Landbouwhgge-school Wageningen 72-10. 342 pp.
117. Maesen, LJG van der. 1979. Genetic resources of grain legumes in the Middle East. In: GC Hawtin & GJ Chancellor (editors). *Food legume improvement and development*: 140–146.
117a. Maeson, LJG van der. 30 December 1981, personal communication.
117c. Marechal, R. 31 August 1981, personal communication.
118. Marshall, DR & P Broue. No date. *The wild relatives of crop plants indigenous to Australia and their use in plant breeding*.
119. Matsuo, T (editor). 1975. *JIBP Synthesis. Volume 5. Gene conservation – exploration, collection, preservation, and utilization of genetic resources*. University of Tokyo Press. 229 pp.
120. Mayr, E. 1982. *The growth of biological thought: diversity, evolution, and inheritance*. The Belknap Press of Harvard University Press. Cambridge (Massachusetts), London (England). 974 pp.
120a. McCollum, GD. 16 July 1981. Personal communication.
121. Meyer, VG. 1974. Interspecific cotton breeding. *Economic Botany* 28: 56–60.
121a. Miller, JD. 16 July 1981, personal communication.
122. Miller, RR. 1977. *IUCN red data book*. Volume 4: Pisces. IUCN. Morges (Switzerland).
122a. Miller, TE. 22 July 1981, personal communication.
123. Mooney, PR. 1979. *Seeds of the earth. A private or public resource?* Canada

Council for International Co-operation.
124. Morgenstern, EK, MJ Holst, AH Teich & CW Yeatman (editors). 1975. *Plus-tree selection: review and outlook.* Canadian Forestry Service Publication No 1347. Ottawa.
125. Mortensen, JA. 1974. Future germplasm reserves in grapes. *Fruit Varieties Journal* 28: 90–94.
126. Muller, CH. 1940. *A revision of the genus Lycopersicon.* USDA Miscellaneous Publication No 382. 29 pp.
127. Nassar, NMA. 1978. Conservation of the genetic resources of cassava (*Manihot esculenta*) determination of wild species localities with emphasis on probable origin. *Economic Botany* 32: 311–320.
128. National Plant Genetic Resources Board. 1979. *Plant genetic resources: conservation and use.* USDA.
129. Nault, LR & WR Findley. In press. *Zea diploperennis: a primitive relative offers new traits for improvement of corn.*
130. Ng, NQ & JT Williams. 1979. *Seed stores for crop genetic conservation.* IBPGR & FAO. AGPE:IBPGR/78/21. Rome. 31 pp.
131. Office of Technology Assessment. 1981. *Impacts of applied genetics: microorganisms, plants, and animals.* Congress of the United States. Washington, DC. 331 pp.
132. Palmberg, C. 1981. Progress in the global programme for improved use of forest genetic resources. In: FAO. *Forest genetic resources information – no 10*: 5–16.
132a. Peries, OS. 10 August 1981, personal communication.
133. Phillips, LL. 1976. Cotton. In: NW Simmonds (editor). *Evolution of crop plants*: 196–200.
133a. Pichel, RJ. 25 May 1983, personal communication.
134. Plucknett, DL, NJH Smith, JT Williams & NM Anishetty, 1983. Crop germplasm conservation and developing countries. *Science* 220: 163–169.
134a. Pochard, E. 1 December 1981, personal communication.
135. Porte, WA & HB Walker. 1945. A cross between *Lycopersicon esculentum* and disease resistant *L. peruvianum. Phytopathology* 35: 931–933.
136. Porter, DM. 1980. *Ms red data bulletin: Galapagos Islands.* Secretariat of IUCN Threatened Plants Committee. Royal Botanic Gardens. Kew.
137. Prescott-Allen, C & R Prescott-Allen. 1986. *The first resource: wild species in the North American economy.* Yale University Press. New Haven & London. 529 pp.
138. Prescott-Allen, R & C Prescott-Allen. 1981. *In situ conservation of crop genetic resources.* FAO/UNEP/IBPGR Technical Conference on Crop Genetic Resources. 6–10 April 1981. Rome.
139. Prescott-Allen, R & C Prescott-Allen. 1982. *What's Wildlife Worth? Economic contributions of wild plants and animals to developing countries.* Earthscan. London. 92 pp.

140. Putt, ED. 1978. History and present world status. In: JF Carter (editor) *Sunflower and science and technology*: 1–29.
141. Quisenberry, KS & LP Reitz (editors). 1967. *Wheat and wheat improvement*. American Society of Agronomy, Inc. Madison. 560 pp.
141a. Rajanaidu, N. 4 March 1982, personal communication.
142. Rick, CM. 1967. Fruit and pedicel characters derived from Galapagos tomatoes. *Economic Botany* 21: 171–184.
143. Rick, CM. 1974. High soluble-solids content in large-fruited tomato lines derived from a wild green-fruited species. *Hilgardia* 42: 493–510.
144. Rick, CM. 1977. Conservation of tomato species germplasm. *California Agriculture* 31: 32–33.
145. Rick, CM. 1979. Potential improvement of tomatoes by controlled introgression of genes from wild species. In: AC Zeven & AM van Harten (editors). *Proceedings of a conference broadening the genetic base of crops*: 167–173.
146. Rick, CM. 1981. *The potential of exotic germplasm for tomato improvement*. Unpublished.
146a. Rick, CM. 15 July 1981, personal communication.
147. Rick, CM, E Kesicki, JF Forbes & M Holle. 1976. Genetic and biosystematic studies on two new sibling species of *Lycopersicon* from interandine Peru. *Theoretical and Applied Genetics* 47: 55–68.
148. Rogers, DJ and SG Appan. 1973. *Flora Neotropica monograph no 13: Manihot Manihotoides (Euphorbiaceae)*. Hafner Press. New York. 272 pp.
149. Rogers, CE, TE Thompson & GJ Seiler. 1982. *Sunflower species of the United States*. National Sunflower Association. Bismarck (North Dakota). 75 pp.
150. Ros, T. 1981. Salmonids in the Lake Vanern area. In: N. Ryman (editor) *Fish gene pools*: 21–31.
151. Ross, H. 1966. The use of wild *Solanum* species in German breeding of the past and today. *American Potato Journal* 43: 63–80.
152. Ross, H. 1979. Wild species and primitive cultivars as ancestors of potato varieties. In: AC Zeven & AM van Harten (editors). *Proceedings of the conference broadening the genetic base of crops*: 237–245.
153. Rowe, PR & DL Richardson. 1975. *Breeding bananas for disease resistance, fruit quality and yield*. Tropical Agriculture Research Services (SIATSA). La Lima, Honduras. Bulletin No 2. 41 pp.
153a. Rowe, PR. 20 August 1981, personal communication.
154. Russell, GE. 1978. *Plant breeding for pest and disease resistance*. Butterworths. London, Boston, 485 pp.
154a. Ruiz, G, Moreno. 11 August 1981, personal communication.
155. Ryman, N (editor). 1981. *Fish gene pools: preservation of genetic resources in relation to wild fish stocks*. Ecological Bulletins No 34. Stockholm (Sweden). 111 pp.

156. Sailer, RI. 1961. Possibilities for genetic improvement of beneficial insects. In: RE Hodgson (editor). *Germ plasm resources*: 295–303.
157. Sakamoto, S. 1976. Breeding of a new sweet potato variety, Minamiyutaka, by the use of wild relatives. *JARQ* 10(4): 183–186.
157a. Sakamoto, S. 18 August 1981, personal communication.
157b. Sakamoto, S. 6 November 1981, personal communication.
158. Sastrapradja, S. 1975. Tropical fruit germplasm in South East Asia. In: JT Williams, CH Lamoureux & N Wulijarni-Soetjipto (editors). *South East Asian plant genetic resources*: 33–46.
159. Siegler, DS (editor). 1977. *Crop resources*. Proceedings of the 17th Annual Meeting of the Society for Economic Botany. 13–17 June 1976. Urbana, Illinois. Academic Press, Inc. New York, San Francisco, London. 233 pp.
160. Simmonds, NW (editor). 1976. *Evolution of crop plants*. Longman. London, New York, 339 pp.
161. Simmonds, NW. 1979. *Principles of crop improvement*. Longman, London, New York. 408 pp.
162. Soria, J. 1975. Recent cocoa collecting expeditions. In: OH Frankel & JG Hawkes (editors). *Crop genetic resources for today and tomorrow*: 175–179.
163. Soule, ME & BA Wilcox (editors). 1980. *Conservation biology: an evolutionary–ecological perspective*. Sinauer Associates, Sunderland (Massachusetts). 395 pp.
164. Spencer, DM (editor). 1978. *The powdery mildews*. Academic Press. London.
165. Sreenivasan, TV, K. Palanichamy & MN Koppar. 1982. Collection of *Saccharum* germplasm in India. *FAO Plant Genetic Resources Newsletter* 52: 20–24.
166. Stegemann, H & V Loeschke. 1979. *Index of European potato varieties: identification by electrophoretic spectra, national registers, appraisal of characteristics, genetic data*. Mitteilungen aus der Biologischen Bundesanstalt fur Land-und Forstwirtschaft, Braunschweig. 233 pp.
167. Stone, BC, JB Lowry, RW Scora & K Jong. 1973. *Citrus halimii*: a new species from Malaya and Peninsular Thailand. *Biotropica* 5: 102–110.
168. Summerfield, RJ & AH Bunting (editors). 1980. *Advances in legume science*. Volume 1 of the Proceedings of the International Legume Conference. 31 July–4 August 1978. Kew Royal Botanic Gardens. Kew. 667 pp.
169. Teich, AH. 1975. Outlook for selected spruces and pines in Canada: white spruce. In: EK Morgenstern, MJ Holst, AH Teich & CW Yeatman (editors). *Plus-tree selection: review and outlook*: 23–33.
170. Thornback, J. 1978. *IUCN red data book*. Volume I: Mammalia. IUCN. Morges (Switzerland).
171. 't Mannetje, L, KF O'Connor & RL Burt. 1980. The use and adaptation of pasture and fodder legumes. In: RJ Summerfield & AH Bunting (editors). *Advances in legume science*: 537–551.

172. Toll, J, NM Anishetty & G Ayad. 1980. *Directory of germplasm collections. 3. Cereals. III. Rice.* IBPGR. AGP:IBPGR/80/109. Rome. 20 pp.
173. Tomes, ML & FW Quackenbush. 1958. Caro-Red, a new provitamin A rich tomato. *Economic Botany* 12: 256–260.
174. USDA. 1981. *Agricultural statistics 1981.* USDA. Washington, DC. 601 pp.
175. Valmayor, RV. 1979. Banana and plantain collections in Indonesia. *FAO Plant Genetic Resources Newsletter* 36: 10–13.
176. Vega, U & KJ Frey. 1980. Transgressive segregation in inter- and intraspecific crosses of barley. *Euphytica* 29: 585–594.
177. Vietmeyer, N. 1981. Man's new best friends. *Quest* Jan/Feb 5(1): 43–49.
178. Vries, CA de. 1974. Sericulture. *Tropical Abstracts* 29: No 9.
179. Watt, WD. No date. *Present and potential effects of acid precipitation on the Atlantic salmon in Eastern Canada*: 39–45.
180. Whyte, RO. 1975. The genetic resources in Asian ecosystems containing perennial species of the Gramineae and the Leguminosae. In: JR Williams, CH Lamoureux & N Wulijarni-Soetjipto (editors). *South East Asian plant genetic resources*: 114–118.
181. Wilkes, HG. 1972. Maize and its wild relatives. *Science* 177: 1071–1077.
182. Wilkes, HG. 1977. Hybridization of maize and teosinte in Mexico and Guatemala and the improvement of maize. *Economic Botany* 31: 254–293.
183. Willan, RL. 1973. Forestry: improving the use of genetic resources. *Span* 16(3): 119–122.
184. Williams, JT, CH Lamoureux & N Wulijarni-Soetjipto (editors). 1975. *South East Asian plant genetic resources*: IBPGR, BIOTROP & LIPI. Bogor. 272 pp.
184a. Wilson, FD. 17 August 1981, personal communication.
185. Winters, HF. 1976. *Microcitrus papuana*, a new species from Papua New Guinea (Rutaceae). *Baileya* 20: 19–24.
186. Wolfe, MS & E Schwarzbach. 1978. The recent history of the evolution of barley powdery mildew in Europe. In: DM Spencer (editor). *The powdery mildews*: 129–157.
187. Wright, JW. 1976. *Introduction to forest genetics.* Academic Press. New York, San Francisco, London. 463 pp.
187a. Yoon, PK. 29 August 1981, personal communication.
188. Zeven, AC & PM Zhukovsky. 1975. *Dictionary of cultivated plants and their centres of diversity.* Pudoc. Wageningen. 219 pp.
189. Zeven, AC & AM van Harten (editors). 1979. *Proceedings of the conference broadening the genetic base of crops.* 3–7 July 1978. Wageningen, Netherlands. Pudoc. Wageningen. 347 pp.
190. Zhukovsky, PM. 1965. Main gene centres of cultivated plants and their wild relatives within the territory of the USSR. *Euphytica* 14: 177–188.
191. Zohary, D. 1970. Centers of diversity and centers of origin. In: OH Frankel & E Bennett (editors). *Genetic resources in plants – their exploration and*

conservation: 33–42.
192. Zohary, D & Z Brick. 1961. *Triticum dicoccoides* in Israel: notes on its distribution, ecology and natural hybridization. *Wheat Information Service* 13: 6–8.
193. McCollum, GD. 16 July 1981, personal communication.

INDEX

Afghanistan, 61, 76, 86
Africa, 22, 25, 32, 35, 38, 61, 81, 84, 93;
  Central, 75;
  East, 35, 63;
  North, 28, 39, 66, 72, 80;
  South, 32, 37, 66;
  West, 24, 36, 37, 39, 46, 68, 70, 74, 75
Akme, 19
Albania, 76
Algeria, 64, 73, 75, 79, 82
All-Union Soya Research Institute, 23
Amazon basin, 35, 36
Americas, 35, 39
Angola, 65
Appan, S.G., 73
Arasu, Dr, 75
Argentina, 29, 32, 63, 64, 65, 66, 67, 87
Asia, 35, 60, 81;
  East, 30;
  Southeast, 17, 23, 35, 36, 60, 61, 76, 92;
  West, 28, 47, 60, 66, 72
Australia, 14, 32, 34, 38, 39, 40, 41, 42, 43, 65, 66, 67, 74, 80, 81, 82, 86, 92
Austria, 30
Avivi, Dr Lydia, 14

Bali, 45
Bangladesh, 17
Bardach, J.E., 47
Batisse, Dr Michel, 91
Baum, Dr Bernard, 72
Beard, James, 27

Beard on Food (Beard), 27
Belgium, 20, 56
Belize, 37, 38
Bislig Bay Lumber Company, 81
Blaak, Ir G., 75
Bockholt, Dr A.J., 18
Bolivia, 40, 64, 65
Brazil, 27, 32, 35, 40, 64, 65, 66, 67, 68, 73, 88, 93
Britain (UK), 19, 20, 31, 86
Bulgaria, 76
Burma, 60

Cameroon, 65
Canada, 31, 38, 48, 63, 65, 83, 86
Canary Islands, 73
Caribbean, 92
Central African Republic, 65
Central America, 27, 71, 92
Centro Internacional de la Papa (CIP; International Potato Centre), 56
Chad, 65
Chile, 63, 64, 65
China, 17, 27, 31, 32, 47, 59, 60, 61, 66, 73, 74, 76, 92
Cocks, P.S., 82
Colombia, 32, 35, 40, 87
Commonwealth Scientific and Industrial Research Organisation (CSIRO), 40, 41
Conway, William G., 88
Costa Rica, 87, 88
Cuba, 32, 67
Cyprus, 42

Czechoslovakia, 86

Dairy Herd Improvement Program, 9
Demuth, R.H., 69
De Wet, Dr Jan, 11

Ecuador, 27, 64, 79, 88
Egypt, 27, 36
England, 20, 41
Ethiopia, 34, 60, 65, 68, 69, 78
Europe, 19, 20, 24, 29, 33, 39, 41, 61, 65
  Eastern, 65
  Western, 82
European Economic Community, 69

FAO. *See* Food and Agriculture Organization
Feldman, Moshe, 16
Fiji, 67, 80
Food and Agriculture Organization (FAO), 38, 67, 68, 69, 94
France, 14, 23, 24, 29, 30, 31, 86
Frankel, Sir Otto, 44–5

Gabon, 23
Galapagos Islands, 54, 76, 79
Germany, East, 86, 87
  West, 19, 21, 30, 86
Global Programme of Improved Use of Forest Genetic Resources, 38
Greece, 27, 41, 42, 73, 76, 86
Guatemala, 71–2
Guzman, Rafael, 18

Harlan, Dr Jack, 11, 27, 61
Harland, S.C., 79
Hawaii, 32, 66
Henzell, Dr E.F., 41
Honduras, 88
Hungary, 47, 86

IBPGR, *See* International Board for Plant Genetic Resources
Iizuka, Dr Muneo, 77
India, 16, 17, 23, 26, 28, 32, 46–7, 57, 60, 64, 65, 66, 67, 68, 76, 86, 88
Indonesia, 16, 17, 24–6, 32, 35, 60, 65, 76, 77, 80
International Board for Plant Genetic Resources (IBPGR), 68, 69, 75, 78, 85, 87, 88, 89, 90, 94
International Institute of Tropical Agriculture (IITA), 22, 23
International Rice Research Institute (IRRI), 17, 93
International Union for Conservation of Nature and Natural Resources (IUCN), 75, 83, 89, 94
Iran, 73, 76
Iraq, 73, 76, 82
Ireland, 20
Israel, 14, 42, 43, 64, 66, 86
Italy, 27, 30, 33, 42, 65, 82, 86
IUCN, *See* International Union for Conservation of Nature and Natural Resources
Ivory Coast, 26, 64

Jamaica, 67, 68
Japan, 23, 24, 61, 66, 74, 77, 86, 92
Java, 25
Jordan, 42, 64

Kenya, 67
Khush, Professor Gurdev, 17
Kinman, Murray, 24
Korea, 61, 74, 92

Lebanon, 64
Leclercq, Patrice, 24
Leon, J., 79
Lesins, Irma, 82
Lesins, Dr Karlis Adolfs, 82

Liberia, 23
Ling, Dr K.C., 17

McLarney, 47
Madagascar, 57, 88
Malawi, 66
Malaysia, 24–6, 35, 75, 77
Mendel, Gregor, 9
Mexico, 17, 27, 32, 37, 40, 61, 64, 65, 71, 73, 93
Miller, Dr J.D., 32
Mongolia, 88
Mooney, Pat, 60
Morocco, 38, 73, 76

Namibia, 65
Nassar, Nagib N.A., 73
Nepal, 17
Netherlands, 21, 75, 86
New Yorker, 27
New Zealand, 59, 63
Niger, 65, 75
Nigeria, 22, 23, 26, 65, 74, 87
North America, 29, 30, 31, 33, 43, 53, 55

Pacific, 81
Pakistan, 32, 76
Papua New Guinea, 65, 67, 77, 80, 88, 93
Paraguay, 65
Peru, 27, 64, 65, 79
Philippines, 16, 17, 32, 37, 67, 80, 81, 87, 88, 93
Phillips, J.R., 82
Poland, 88
Portugal, 42, 43

Rajanaidu, Dr, 75
Rick, Dr Charles, 76
Rogers, David J., 73
Romania, 27, 30
Ryder, 47

Salim, Dr Emil, 60
Sears, Ernest R., 16
Senegal, 57
Seychelles, 23
Sharma, Dr S.D., 17
Sierra Leone, 23
Somalia, 65
Soule, Dr Michael, 44–5
South Africa, 32, 37, 66
South America, 20, 23, 24, 34, 36, 53, 60, 61, 73, 75, 81
Spain, 20, 27, 33, 41
Sri Lanka, 16, 34, 35, 89
Sudan, 34, 36, 37, 65, 75
Sumerians, 47
Sweden, 83, 87
Syria, 64, 73, 76
Taiwan, 73, 74
Tanzania, 23, 66
Tapajos, 35
Thailand, 32, 77
't Mannetje, Dr L., 41
Trinidad, 88
Tunisia, 42, 64
Turkey, 27, 33, 64, 73, 76, 86

Uganda, 40, 57, 84
United Kingdom (UK). *See* Britain
United Nations, 38;
    Conference on the Human Environment, 90;
    UNEP (United Nations Environment Programme), 68, 94;
    UNESCO (United Nations Educational, Scientific and Cultural Organization), 91, 94
Uruguay, 65, 68
USA, 9, 14, 17, 23, 27, 29, 32, 34, 36, 39, 56, 57, 61, 63, 64, 65, 66, 67, 84, 86, 87, 88, 93
USSR, 14, 20, 23, 28, 30, 31, 34, 36, 47, 58, 60–1, 65, 66, 76, 86, 87, 89, 93

Vavilov, 60–1
Venezuela, 64, 88
Vietnam, 16

Weizmann Institute of Science, 14
Whiteside, Thomas, 27
Wickham, Henry, 35
Williams, Dr J.T., 69

Yemen, 34
Yugoslavia, 41

Zaire, 23, 57, 64
Zanzibar, 48
Zhukovsky, P.M., 61
Zimbabwe, 34, 38, 67

Robert and Christine Prescott-Allen are resource analysts and writers, specializing in conservation and development of plant and animal species and gene pools.

For Product Safety Concerns and Information please contact our
EU representative GPSR@taylorandfrancis.com Taylor & Francis
Verlag GmbH, Kaufingerstraße 24, 80331 München, Germany